The Leukemias: Epidemiologic Aspects

Monographs in Epidemiology and Biostatistics
edited by Abraham M. Lilienfeld

1 THE EPIDEMIOLOGY OF DEMENTIA
James A. Mortimer and Leonard M. Schuman
1981

2 CASE CONTROL STUDIES:
Design, Conduct, Analysis
James J. Schlesselman
1982

3 EPIDEMIOLOGY OF MUSCULOSKELETAL DISORDERS
Jennifer L. Kelsey
1982

4 URBANIZATION AND CANCER MORTALITY:
The United States Experience, 1950–1975
Michael R. Greenberg
1983

5 AN INTRODUCTION TO EPIDEMIOLOGIC METHODS
Harold A. Kahn
1983

6 THE LEUKEMIAS: EPIDEMIOLOGIC ASPECTS
Martha S. Linet
1984

Monographs in Epidemiology and Biostatistics
Volume 6

The Leukemias: Epidemiologic Aspects

Martha S. Linet, M.D., M.P.H.

SCHOOL OF HYGIENE AND PUBLIC HEALTH
THE JOHNS HOPKINS UNIVERSITY

New York Oxford
OXFORD UNIVERSITY PRESS
1985

Library of Congress Cataloging in Publication Data

Linet, Martha S.
The leukemias : epidemiologic aspects.

(Monographs in epidemiology and biostatics ; v. 6)
Bibliography: p.
Includes index.
1. Leukemia—Addresses, essays, lectures.
2. Leukemia—Etiology—Addresses, essays, lectures.
3. Epidemiology—Addresses, essays, lectures. I. Title.
II. Series. [DNLM: 1. Leukemia—Occurrence. 2. Leukemia
—Mortality. W1 M0567LT v.6 / WH 250 L754L]
RC643.L483 1984 616.99'419071 83-26783
ISBN 0-19-503448-1

Printed in the United States of America

Printing (last digit): 9 8 7 6 5 4 3 2 1

*To my loving husband Paul, our daughter Alice Suzanne,
and with special thanks to my parents*

Preface

Despite their rarity, the leukemias have received extensive epidemiologic study. Both descriptive and analytic studies suggest that the leukemias comprise a heterogeneous group of disorders with different sociodemographic features and risk factors characterizing the histopathologic and clinical subtypes. A number of papers have reviewed the literature, but there is no comprehensive summary of the epidemiologic literature organizing and critically evaluating important studies of the leukemias by cell type.

Only a limited amount of research in this area has focused on the leukemias individually; much of the literature, particularly earlier studies, considered the leukemias as a single entity. The latter types of studies have been included in this book, although, whenever possible, data are presented by cell type.

Over the past few years a number of classification systems have been used for diagnostic and prognostic purposes. Epidemiologic findings suggest that subclassification may be useful in describing etiologic differences as well. Classification methods are continuing to evolve as more knowledge is gained from study of immunologic characteristics (e.g. cell surface markers), chromosomal aspects, and biochemical features. Epidemiology may also provide useful information for development of more precise classification schemes. For this monograph the leukemias have been subdivided into four major groups: acute lymphocytic (ALL), chronic lymphocytic (CLL), acute nonlymphocytic (ANLL), and chronic myelocytic leukemia (CML). As epidemiologic research continues to focus separately on the individual leukemias and as knowledge from many disciplines continues to increase, revision will certainly be necessary.

This book should prove useful not only to epidemiologists and other public health scientists but also to preventive medicine practitioners, oncologists, hematologists, other clinicians, environmental health specialists, scientists in regulatory agencies, and others concerned with issues related to potential leukemogenicity.

It is also important to recognize that, in spite of extensive research, relatively little is known about the etiology of a substantial proportion of each of the leukemias. It is hoped that the approach described, which focuses separately on each individual, relatively homogeneous subtype of this group of heterogeneous disorders, will advance understanding in future epidemiologic studies. Although important progress in treatment has been made for some of the leukemias, most therapies have significant side effects and are quite costly. Therefore it is crucial not to lose sight of the importance of preventing—if possible—development of these serious malignancies. This book is intended to stimulate additional studies in this area and thereby identify etiologic factors (such as radiation) that can be reduced or possibly eliminated.

June 1984 M.S.L.
Baltimore, Md.

Acknowledgments

I am deeply indebted to the late Dr. Abraham M. Lilienfeld whose warm enthusiastic support and editorial guidance were instrumental in the completion of this monograph. I have greatly benefited from his encouragement in this project, which grew out of his perception of the need for an updated review of the leukemias by cell-specific type.

Collaboration with Dr. Moyses Szklo in population-based case-control epidemiologic studies of chronic lymphocytic and acute nonlymphocytic leukemia has been an important source of stimulation for my continuing interest in all the leukemias. I am also very grateful for the challenging intellectual environment he continuously provides.

Dr. Genevieve Matanoski graciously reviewed the section on radiation and made many useful suggestions.

I am especially appreciative of the major editorial assistance provided by Monica Kinosian during all phases of the preparation of this monograph. Her interest in this difficult topic and unflagging enthusiasm helped see this project through.

Many thanks are also due to Ruth Barnes and Alverta Fields for superb typing and preparation of tables.

This work was supported in part by Public Health Service Grant Number CA26500 awarded by the National Cancer Institute.

Abraham M. Lilienfeld, M.D., M.P.H., D.Sc. (Hon.)
In Memoriam

On August 6, 1984, the editor of Monographs in Biostatistics and Epidemiology, one of the most recognized and distinguished epidemiologists of the twentieth century, passed away. Abe Lilienfeld's pioneering work in reproductive, cardiovascular, and cancer epidemiology encompasses the key substantive as well as methodologic issues in all these areas. In addition to focusing on etiologic factors, Abe was deeply concerned with the public health consequences of his research.

He was responsible for expanding the epidemiologic data collection program at the Roswell Park Memorial Institute that became a model resource for many important case control cancer studies. His participation in multicenter collaborative epidemiologic studies not only resulted in many important findings and contributions to public health, but demonstrated as well his unique ability to overcome the difficulties of working productively with several collaborators.

More recently he simultaneously pursued research projects in digestive diseases and the history of epidemiology and edited a number of volumes in this and other series of monographs and reviews.

Dr. Lilienfeld's positions throughout his illustrious career included Chairman of the Department of Chronic Diseases and its successor, the Department of Epidemiology of the Johns Hopkins School of Hygiene and Public Health; Staff Director of the National Commission on Heart Disease, Cancer and Stroke, and principal author of the Commission's report (forming the basis for the Johnson Administration's legislative proposals establishing national regional medical programs); founder, together with Brian MacMahon and Milton Terris, of the Society for Epidemiologic Research and one of the chief organizers of the American College of Epidemiology. He was the recipient of the Bronfman Award for Achievement in Public Health (1968), was elected to membership in the Institute of Medicine of the National Academy of Sciences (1975), received an Honorary Doctor of Science degree from the University of Maryland School of Medicine (1975) and the John Snow Award from the American Public Health Association (1975); he was named University Distinguished Service Professor of Epidemiology by The Johns Hopkins University (1975).

More important perhaps than all of these illustrious and remarkable achievements and awards was the far-reaching effect he had on his students, colleagues and friends. His warm encouragement and suggestions generously given to all will long be remembered and will continue to permeate epidemiologic studies.

Contents

The Leukemias: Epidemiologic Aspects

1 Heterogeneity of the Leukemias

Earlier epidemiologic reviews considered leukemia as a single disease with some minor differences in risk factors for each of the histopathologic types [1–6]. In the last two decades, epidemiologic studies have provided evidence that the diseases collectively identified as leukemia are several distinct malignancies characterized by varying age, sex, race, and ethnic group patterns, dissimilar secular trends, and different etiologic factors. This evidence substantiates the diversity in the leukemias first noted by pathologists in the latter half of the nineteenth and early twentieth centuries and later described by hematologists.

Leukemia was probably first described clinically by Velpeau in 1827 [7, 8] but it was not until 1847 that Virchow named the disease "leukemia" [9]. By 1889 acute and chronic leukemia were recognized as two separate entities [10]. In 1900 Naegeli described the myeloblast cell and separated acute leukemia into two categories: myelocytic and lymphocytic [11]. With the introduction of the peroxidase stain in 1916 and the development of additional cytochemical techniques, acute myelocytic leukemia was further divided into subtypes: monocytic, myelomonocytic, and promyelocytic [7, 8].

CLASSIFICATION

Over the years, clinicians have developed a variety of classification schemes to determine prognosis and to define subgroups of leukemia patients who might benefit from particular therapies; a reproducible classification can provide a firm basis for comparing different drug regimens in similar groups of patients [8]. A recently proposed classification for *acute lymphocytic leukemia* (ALL) defines four subtypes, null cell ALL, common ALL, T-cell ALL and B-cell ALL, which appear to have somewhat different sex distributions, prognoses, and responses to treatment [12, 13].

A group of seven French, American, and British hematologists (FAB group) initially studied peripheral blood and bone marrow films from 200 cases of *acute nonlymphocytic leukemia* (ANLL) in 1974–75 and, working together and independently, proposed the following six categories (designated M1–M6): acute undifferentiated leukemia, promyelocytic leukemia, acute myelocytic leukemia with Auer rods, myelomonocytic leukemia, monocytic leukemia (subclassified into a monoblastic and a more differentiated monocytic type), and erythroleukemia. This scheme is only applied to overt cases of ANLL, with greater than 50% blasts and promyelocytes present in the bone marrow [14, 15].

The Chronic Leukemia-Myeloma Task Force of the National Cancer Institute has suggested specific diagnostic criteria for dividing *chronic lymphocytic leukemia* (CLL) into two subtypes: indolent and rapidly progressive. This Task Force also proposed criteria for defining *chronic myelocytic leukemia* (CML) and divided all cases into two subtypes, based on the absence or presence of the Philadelphia (Ph^1) chromosome, an acquired abnormality [16].

Other clinical classifications have attempted to differentiate the stages of disease within a particular subtype of leukemia. Rai et al. recently described a clinical staging system for CLL which appears to accurately predict survival but has not been extensively evaluated [17]. A method for staging CML was first reported in July 1980; it also requires additional study [18].

Cytogenetic methods have been used to identify two major categories of CML, those with and those without the Philadelphia chromosome (Ph^1), but additional chromosomal abnormalities have been described [19–25]. Chromosomal abnormalities have also been observed in a significant proportion of acute nonlymphocytic leukemia cases (ANLL) [24–29] and acute lymphocytic leukemia (ALL) cases [30]. Chromosomal breaks, duplications, translocations, or other types of abnormalities have rarely been associated with chronic lymphocytic leukemia [31].

Another classification system that appears promising in predicting survival is an immunologic grouping of the lymphocytic leukemias by lymphocyte cell type, characterizing the lymphocytes as B-cells, T-cells or null cells [13, 32–35]. This classification may have great potential in determining prognosis for ALL.

THE FOUR HISTOPATHOLOGICAL TYPES

Epidemiologic Rationale

Most epidemiologic studies of leukemia have failed to separate different subtypes of cases, in part because they have not recognized epidemiologic differences between these types. The rarity of leukemia of all types combined and the even lower incidence of cases in each separate subtype make it difficult to obtain adequate sample sizes for studies. Given the recognition of differences between subtypes of cases, an important issue is how to define these subtypes to reveal potential differences in etiology, prognosis, and treatment. Until recently, most subtypes have been characterized on the basis of histopathologic and/or clinical differences. This method has proven useful for classifying leukemia patients into subtypes that have repeatedly and reliably displayed differences. A combined histopathologic and clinical classification scheme appears to be appropriate for defining groups that deserve separate study in epidemiologic research.

Major epidemiologic differences between the four histopathologic subtypes are described in greater detail in this monograph. Age distributions vary considerably. ALL is primarily a disease of the first two decades of life. It is uncommon between ages 25 and 60 but increases in incidence with age particularly after 60. CLL is extremely rare before age 40 but increases in incidence with age, especially after age 50. Approximately 30% of acute childhood leukemias are acute nonlymphocytic leukemias (ANLL) (which include acute myelocytic leukemia or AML, acute monocytic leukemia or AMoL, acute myelomonocytic leukemia or AMMoL, acute promyelocytic leukemia or APL, and other very rare leukemias). ANLL increases in incidence with age; over 50% of adulthood leukemia is ANLL. CML, also called chronic granulocytic leukemia (CGL), is highest in incidence in those aged 30–60 but a few cases occur in children and adolescents.

ALL, ANLL, and CML have a slightly higher incidence rate in males compared with females, with a male/female ratio of 1.1–1.2. However, the male to female ratio for CLL is higher, 1.7.

The incidence of all leukemias at all ages is consistently higher in whites than nonwhites. CLL is also very infrequent in Asians, even in second and third generation migrants to western countries.

Survival patterns appear to differ considerably among subtypes. However, recent studies evaluating survival must be interpreted with some caution because results may be highly affected by rapidly changing and improved therapies, especially in the last few years. In general, children and adults with untreated ALL and ANLL have a very short survival—a few weeks to a few months. However, for patients with ALL receiving the current therapy of combination drug induction and maintenance regimens, the median survival without evidence of disease is 60 months or more. A significant proportion of children with this condition have had very long survivals, years to decades, and are considered cured. More than 50% of persons with ANLL treated with aggressive multidrug chemotherapy achieve a remission although it is frequently of a fairly short duration. Persons with CLL have a median survival of approximately 3 years. How-

Table 1-1 Classification of the leukemias

Major morphologic types and variants	Primary involved age group(s)	Median survival	Characteristics of the leukemia cell involved
Acute lymphocytic leukemia (ALL)	Children	24 to > 48 months (depending on treatment given, response obtained and treatment center)	Most common: null-cell precursors of lymphocytes Much less common: T-cell precursors Rare: B-cell precursors
Chronic lymphocytic leukemia (CLL)	The elderly	3–4 years	Most common: a monoclonal disorder of B lymphocytes Rare: T-cell CLL
Acute nonlymphocytic leukemia (ANLL) Most common: Acute myelocytic leukemia (AML) Other variants: Acute monocytic leukemia Myelomonocytic leukemia Promyelocytic leukemia Undifferentiated acute leukemia of adults	Chiefly adults Less common in children	10–14 months	Leukemic cells probably arise from precusors of myeloid elements or even earlier common stem cells
Chronic myelocytic leukemia (CML)	Middle-aged adults	3–4 years	Leukemic cells arise from a single abnormal clone that is a precursor of myeloid, erythroid and megakaryocytic cells but not lymphocytes. The abnormal cells contain the Ph^1 chromosome in 90% of the cases

Source: Cline and Haskell [36].

ever, a significant proportion of cases (10%–20%) are completely asymptomatic at the time of diagnosis. The condition may have been discovered by accident in the course of a work-up for another, unrelated condition. The response of patients with CLL to therapy is not striking, and therapy in general does not affect survival. The indolent subtype of CLL cases has a long survival, with death usually from an unrelated disease. CML patients have chronic leukemia for a median duration of 3 years. It then transforms into a blastic phase in which the disease appears similar to an acute leukemia and is often unresponsive to chemotherapy [31].

There are several risk-factor differences between the histopathologic types [1–6]. CLL appears to have a strong genetic component; approximately 10% of first-degree relatives of cases have been found to have CLL or another type of lymphoproliferative malignancy. While more than one case of other types of leukemia in a family have been reported clinically, the genetic component is not as strong as in CLL.

Radiation can be an etiologic factor in all types of leukemia except CLL. However, it appears that different types of radiation can cause histopathologically different types of leukemia. Evidence from atomic-bomb survivors in Japan substantiates this. The incidence of CML was high in Hiroshima where both gamma rays and neutrons were produced by the A-bomb, but much lower in Nagasaki, where the radiation was primarily from gamma rays. Benzene has been primarily associated with ANLL; follow-up studies of rubber workers, presumed to have a high exposure to benzene, also revealed higher than expected incidence of CLL.

Various types of chemotherapy (primarily alkylating agents but other types as well) have caused higher incidences of ANLL.

Infectious agents have been shown to have only a possible weak association with development of ALL in childhood and rare T-cell leukemias of Japanese and West Indian adults.

Given these differences and the small number of studies that have examined leukemia histopathologically, it seems appropriate to consider four major groupings of the leukemias for the purposes of this monograph. A summary of the characteristics of the classification scheme used is provided in Table 1-1.

Clinical Description

ALL Acute lymphocytic leukemia is a hematopoietic malignancy involving progressive infiltration and replacement of normal bone marrow and lymphatic tissue by an abnormal lymphoid precursor with immunocyte-specific determinants. This precursor is often identified as a "null" cell but in a small number of cases is of T-cell, or, even more rarely, B-cell origin [34, 35]. The malignancy occurs predominantly in children and is characterized by systemic complaints of malaise, fatigue, weight loss, pallor, easy bruising or bleeding; moderate lymphadenopathy, splenomegaly and hepatomegaly are common. Bony pain and joint swelling may occur [8, 31, 36].

The pathogenesis of ALL is somewhat obscure and the reason(s) for the high incidence in children 2–9 years old is not known. As the disease develops, malfunctioning primitive cells appear to accumulate in blood-forming tissue; the leukemia cells appear to replicate at the same rate or even slower than comparable normal lymphoblasts.

Before or simultaneously with leukemic cell replication, normal hematopoiesis becomes impaired. The mechanism of this is unclear. Proposed explanations include failure in the production or expression of humoral regulators of the blood cell system and inhibition of normal cell maturation by leukemic cells. Whether ALL is a monoclonal disease is unknown. Whether the disease consists of a defect involving one or more types of replicating blood cells is also unclear. Currently, ALL appears to be a disease in which normal and malignant cells coexist and compete within the blood-forming organs. While epidemiologic studies may point to associations between various factors and the onset of ALL, further understanding of the mechanisms by which this may occur will require more knowledge of the basic pathophysiology of the disease, including the following:

1. Identification of the primary level of leukemia involvement (i.e., the clone(s) involved at the onset of the disease), which may require additional morphologic, biochemical, and immunologic approaches to the identification of leukemic cells.
2. The mechanism responsible for the ineffective hematopoiesis.
3. Identification of microenvironmental factors that may play a key role in disease onset, remission, and relapse.

CLL Chronic lymphocytic leukemia is a lymphoproliferative or progressive abnormality of the lymphoid tissues primarily affecting the small B lymphocytes. Clinically, patients may present with lymphadenopathy, splenomegaly, abnormal lymphocytes in peripheral blood, infiltration of bone marrow, skin, other organs, and sometimes systemic symptoms such as weight loss, malaise, fever. The disease may vary considerably in severity and may range in course from rapidly progressive with development of anemia, thrombocytopenia, significant organ infiltration, infection to completely asymptomatic. Median survival following diagnosis is 3–4 years; but 10%–15% of patients survive as long as 10–15 years with little or no treatment [8, 31]. The disease virtually never occurs in the young; the incidence rate rapidly rises in late middle age and is highest in the oldest age groups [37].

The onset of CLL is marked by the development of an abnormal clone of leukemic cells. It is not clear where the abnormal lymphocytes originate.

Most of the cells are long-lived, have little proliferative capacity, and are not active immunologically. The mechanisms involved in the pathogenesis of the disease are unclear. A significant number of families with an index case of CLL have two or more members with CLL, other lymphoproliferative abnormalities, or abnormalities in immunologic function [8, 31].

ANLL The various acute nonlymphocytic leukemias, including acute myelocytic leukemia, monocytic leukemia, myelomonocytic leukemia, acute promyelocytic leukemia and others, share many clinical characteristics and are very similar in their descriptive epidemiologic patterns [4, 8, 37]. Each subtype of ANLL will first be described separately and then all the subtypes will be considered as a group. Usually ANLL is a rapidly progressive disease, although the neoplastic cells grow at varying rates. Prior to the introduction of chemotherapy in the 1940's, median survival was less than 6 months.

Many of the acute nonlymphocytic leukemias are believed to arise from precursors of myeloid elements or even earlier common stem cells, though other theories have been postulated. Identical chromosome abnormalities have been noted in the erythroid and in the myeloid precursor cells in acute myelocytic leukemia. Many of the myeloproliferative disorders appear to be interrelated; the initial neoplastic change in many of these related hematopoietic malignancies is postulated to occur in a pluri-potential marrow stem cell. The various types of ANLL differ morphologically and histochemically despite a probable common precursor origin. The risk factors for all the myeloproliferative malignancies appear to be similar. How morphologically and cytochemically different leukemias can result from a probable common precursor and similar risk factors is unclear.

Acute myelocytic leukemia is a hematopoietic malignancy of the granulocyte precursors. During the early phases of the disease, the myeloid elements proliferate rapidly and the malignant cells accumulate in bone marrow. Eventually normal hematopoiesis fails and clinical symptoms appear which are similar to those present in ALL. Lymphadenopathy occurs less often in AML than in ALL, while splenomegaly and hepatomegaly are quite common. In general, the disease has a rapidly progressive course. AML constitutes a significant percentage of acute leukemias occurring in the first few months of life but the incidence rate is fairly low in individuals aged 1–20 years. From age 20 years onward, the incidence rate increases progressively [8, 31].

Acute monocytic and *acute myelomonocytic leukemia* are clinically very similar to AML except for greater frequency of skin, mucous membrane, gum and other extramedullary involvement. Abnormal monocytes invade blood, bone marrow, and other tissues [8, 31].

Acute promyelocytic leukemia also shares many characteristics with the other nonlymphocytic leukemias, though it tends to display disseminated intravascular coagulation, a condition in which there can be extensive bleeding due to pathologic consumption of clotting factors [8, 31].

The pathogenesis of the various subtypes of ANLL is not well understood. However, some risk factors for ANLL have been identified and the pathophysiologic mechanisms are being investigated [38]. In this monograph, ANLL and AML have sometimes been used interchangeably because much of the data are available only for AML.

CML Chronic myelocytic leukemia is a neoplastic disease with excessive proliferation of granulocytes in bone marrow, and extramedullary myelopoiesis in the spleen and liver. Large numbers of immature granulocytes circulate in peripheral blood. CML is associated with a unique chromosome abnormality in 90% of cases [8, 31, 39]. The earliest symptoms are usually systemic and include malaise, fatigue, pallor, and sweating. Bony pain, especially of the sternum, and abdominal discomfort, secondary to splenomegaly, may occur. In the early stages, the disease can often be controlled and may not progress for 1–4 years. After this period, disease transition and progression may occur and CML may take on the characteristics of an acute nonlymphocytic leukemia unresponsive to chemotherapy.

The virtually pathognomonic chromosome (an acquired and not a congenital abnormality) and an abnormal leukemia cell strain probably appear following the leukemogenic effects of some types of ionizing radiation and other risk factors for CML [39]. Studies of bone marrow have shown that the myeloid, megakaryocytic, and probably erythroid cell strains contain the abnormal chromosome but that the lymphocytes of such individuals do not contain the Ph^1 abnormality [40]. Further evidence that the Ph^1 abnormality is acquired comes from twin studies. When one of a pair of twins developed CML, both were studied; the abnormal chromosome was present only in the affected twin [4, 7].

Theories on the clonal nature of the disease are supported by two types of evidence: confinement of the chromosomal abnormality to four types of hematopoietic cells (granulocyte, erythrocyte, platelet, and monocyte/macrophage), all believed to derive from a common stem cell, and from study of the heterozygotes for the X-linked glucose 6-phosphate dehydrogenase locus [40]. Reasons for the apparent selection and growth advantage of the abnormal leukemic cells over the normal hematopoietic cells are not clear.

REFERENCES

1. J. Clemmesen, "Epidemiological aspects of human leukemia," in *International Wenner-Gren Symposium,* pp. 161–177, (E. Winqvist, ed). Oxford: Pergamon Press, 1966.

2. I.I. Kessler and A.M. Lilienfeld, "Perspectives in the epidemiology of leukemia," *Adv. Cancer Res.* **12**:225–302 (1969).
3. J. Clemmesen, "On the epidemiology of leukemia," in *Advances in Acute Leukemia,* pp. 1–50 (F.J. Cleton, D. Crowther, and J.S. Malpas, eds). Amsterdam: North-Holland; New York: American Elsevier, 974.
4. C.W. Heath, Jr., "The epidemiology of leukemia," in *Cancer Epidemiology and Prevention: Current Concepts,* pp. 318–350 (D. Schottenfeld, ed). Springfield, Illinois: Charles C. Thomas, 1975.
5. M. Alderson, "The epidemiology of leukemia," *Adv. Cancer Res.* **31**:1–76 (1980).
6. C.W. Heath, Jr., "The leukemias," in *Cancer Epidemiology and Prevention,"* pp. 728–738 (D. Schottenfeld and J.F. Fraumeni, Jr., eds). Philadelphia: W.B. Saunders, 1982.
7. F. Gunz and A.G. Baikie, "Leukemia in the past," in *Leukemia,* 3rd edn, pp. 3–12. New York: Grune & Stratton, 1974.
8. M.M. Wintrobe, G.R. Lee, D.R. Boggs, T.C. Bithell, J.W. Athens, and J. Foerster, "Classification, pathogenesis, and etiology of neoplastic diseases of the hematopoietic system," in *Clinical Hematology,* 8th ed, pp. 1432–1471, Philadelphia: Lea & Febiger, 1981.
9. R. Virchow, "Die leukämie," in *Gesammelte Abhandlungen zur Wissen Schaftlichen Medizin,* pp. 190–211. Frankfurt: Meidinger, 1865.
10. W. Ebstein, "Ueber die acute leukämie and pseudoleukämie," *Deutsches Arch. Klin. Med.* **44**:343–396 (1889).
11. O. Naegeli, "Ueber rothes knochenmark und myeloblasten," *Deutsch Med. Wochenschr.* **26**:287–290 (1900).
12. G. Mathé, P. Pouillart, R. Weiner, M. Hayat, M. Steresco and M. Lafluer, "Classification and subclassification of acute leukemias correlated with clinical expression, therapeutic sensitivity and prognosis," *Recent Results Cancer Res.* **43**:6–20 (1973).
13. J.M. Chessells, R.M. Hardisty and M.F. Greaves, "Acute lymphoblastic leukemia in children: classification and prognosis," *Lancet* **2**:1307–1309 (1977).
14. H.R. Gralnick, D.A.G. Galton, D. Catovsky, C. Sultan and J.M. Bennett, "Classification of acute leukemia," *Ann. Intern. Med.* **87**:740–753 (1977).
15. French-American-British (FAB) Co-operative Group: J.M. Bennett, D. Catovsky, M-T. Daniel, G. Flandrin, D.A.G. Galton, H.R. Gralnick and C. Sultan, "Proposals for the classification of the acute leukemias," *Br. J. Haematol.* **33**:451–458 (1976).
16. Chronic Leukemia-Myeloma Task Force, "Proposed guidelines for protocol studies. III. Chronic lymphocytic leukemia and IV. New agents in the treatment of chronic granulocytic leukemia," *Cancer Chem. Rep.* (Part 3) 41–62 (1968).

17. K.R. Rai, A. Sawitsky, E.P. Cronkite, A.D. Chanana, R.N. Levy and B.S. Pasternack, "Clinical staging of chronic lymphocytic leukemia," *Blood* **46:**219–234 (1975).
18. S. Tura, M. Baccarani and G. Corbelli, "Staging of chronic myeloid leukaemia," *Br. J. Haematol.* **47:**105–119 (1981).
19. G. Gahrton, J. Lindsten and L. Zech, "Involvement of chromosomes 8, 9, 19 and 22 in Ph^1 positive and Ph^1 negative chronic myelocytic leukemia in the chronic or blastic stage," *Acta Med. Scand.* **196:**355–360 (1974).
20. S.I. Sonta and A.A. Sandberg, "Chromosomes and causation of human cancer and leukemia. XXIV. Unusual and complex Ph^1 translocations and their clinical significance," *Blood* **50:**691–697 (1977).
21. N.J. Shiffman, E. Stecker, P.E. Conen and H.A. Gardner, "Males with chronic myeloid leukemia and the 45, XO, Ph^1 chromosome pattern," *Can. Med. Assoc. J.* **110:**1151–1154 (1974).
22. T. Mohandas, C. Anderson and D. Okun, "Chronic myelogenous leukemia with a complex translocation," *Cancer Genet. Cytogenet.* **2:** 19–21 (1980).
23. S. Kohno, S. Abe and A.A. Sandberg, "The chromosomes and causation of human cancer and leukemia. XXXVIII. Cytogenetic experience in Ph^1-negative chronic myelocytic leukemia (CML)," *Am. J. Hematol.* **7:**281–291 (1979).
24. M. Sakurai and A.A. Sandberg, "The chromosomes and causation of human cancer and leukemia. XVIII. The missing Y in acute myeloblastic leukemia (AML) and Ph^1-positive chronic myelocytic leukemia (CML)," *Cancer* **38:**762–769 (1976).
25. A.W. Wayne, J.C. Sharp, M.V. Joyner, H. Sterndale and K.A. Pulford, "The significance of Ph^1 mosaicism: a report of six cases of chronic granulocytic leukaemia and two cases of acute myeloid leukaemia," *Br. J. Haematol.* **43:**353–360 (1979).
26. S. Abe, H.M. Golomb, J.D. Rowley, F. Mitelman and A.A. Sandberg, "Chromosomes and causation of human cancer and leukemia. XXXV. The missing Y in acute non-lymphocytic leukemia (ANLL)." *Cancer* **45:**84–90 (1980).
27. H.M. Golomb, J.W. Vardiman, J.D. Rowley, J.R. Test and V. Mintz, "Correlation of clinical findings with quinacrine-banded chromosomes in 90 adults with acute nonlymphocytic leukemia: an eight-year study (1970–1977)," *N. Engl. J. Med* **299:**613–619 (1978).
28. M. Sakurai and A.A. Sandberg, "Chromosomes and causation of human cancer and leukemia. IX. Prognostic and therapeutic value of chromosomal findings in acute myeloblastic leukemia," *Cancer* **33:**1548–1557 (1974).
29. J.J. Yunis, C.D. Bloomfield and K. Ensrud, "All patients with acute

nonlymphocytic leukemia may have a chromosomal defect," *N. Engl. J. Med.* **305:**135–139 (1981).
30. J.R. Priest, L.L. Robinson, R.W. McKenna, L.L. Linquist, P.I. Warkentin, T.W. Lebien, W.G. Woods, J.H. Kersey, P.F. Coccia, and M.E. Nesbit, Jr., "Philadelphia chromosome-positive childhood acute lymphoblastic leukemia," *Blood* **56:**15–22 (1980).
31. W.J. Williams, *Hematology,* 2nd ed. New York: McGraw-Hill, 1977.
32. E.L. Reinherz, L.M. Nadler, S.E. Sallan and S.F. Schlossman, "Subset derivation of T-cell acute lymphoblastic leukemia in man," *J. Clin. Invest.* **64:**392–397 (1979).
33. J.D. Bitran, "Prognostic value of immunologic markers in adults with acute lymphoblastic leukemia," *N. Engl. J. Med.* (Letter to the Editor) **299:**1317 (1978).
34. L. Sen and L. Borella, "Clinical importance of lymphoblasts with T markers in childhood acute leukemia," *N. Engl. J. Med.* **292:** 828–832 (1975).
35. S.E. Sallan, L. Chess, E. Frei III, C. O'Brien, D.G. Nathan, J.L. Strominger and S.F. Schlossman, "Utility of B- and T-cell specific antisera in the classification of human leukemias," in *Differentiation of Normal and Neoplastic Hematopoietic Cells,* pp. 479–484 (B. Clarkson, P. Marks and J. Till, eds). Cold Spring Harbor: Cold Spring Harbor Laboratory, 1978.
36. M.J. Cline and C.M. Haskell, *Acute Leukemia and Chronic Leukemia in Cancer Chemotherapy,* 3rd ed, pp. 242–286. Philadelphia: W.B. Saunders, 1980.
37. S.J. Cutler and J.L. Young, (eds) *Third National Cancer Survey: Incidence Data,* National Cancer Institute Monograph No. 41. Washington, D.C.: U.S. Government Printing Office, 1975.
38. M. Szklo, "Aplastic anemia," in *Reviews in Cancer Epidemiology,* Volume 1, pp. 218–244 (A.M. Lilienfeld, ed). New York: Elsevier/North Holland, 1980.
39. H.P. Koeffler and D.W. Golde, "Chronic myelogenous leukemia—new concepts," *N. Engl. J. Med.* **304:**1201–1209, 1269–1274 (1981).
40. P.J. Fialkow, R.J. Jacobson and T. Papayannopoulou, "Chronic myelocytic leukemia: clonal origin in a stem cell common to the granulocyte, erythrocyte, platelet and monocyte/macrophage," *Am. J. Med.* **63:**125–130 (1977).

2 Methodologic Issues

Many important methodologic issues must be considered when interpreting the results of epidemiologic studies of the leukemias. General methodologic problems are discussed in this chapter. In individual sections that follow, methodologic problems which relate to a specific condition or risk factor are reviewed.

The classification of leukemias adopted in this book is based on histopathologic and clinical groupings familiar to morphologists, clinicians and, more recently, some epidemiologists, though many previous studies examined leukemia as a single entity or combined one or more of the histopathologic groups. We have attempted to find group-specific histopathologic data within such studies. A lack of group-specific information did not preclude the use of a study if the collective results could be interpreted with our approach. The specific subtypes defined in this book are based on older morphologic schemes that have been widely used over the past 15–30 years. As techniques for lymphocyte typing, high resolution chromosome analysis, etc., become common in the clinical setting, new immunologic, cytogenetic, and other classifications may find widespread use. Epidemiologists should work alongside clinicians, pathologists, geneticists, and immunologists to develop, evaluate, and apply classification systems that will hasten understanding of the etiologies of the leukemias.

Major alterations were made in the leukemia classification in the 7th, 8th, and 9th Revisions of the International Classification of Diseases Coding System. In the 6th Revision (adopted in 1948), leukemia and aleukemia were combined in a single rubric. No distinction was made between lymphocytic and myelocytic leukemia or between acute and chronic leukemia. The 7th Revision (adopted in 1958) designated five categories for leukemia; separate code numbers were designated for lymphatic, myeloid, monocytic, acute, and other and unspecified leukemia. In retrospect, it is unclear why acute leukemia should be labeled separately, rather than as a form of lymphatic, myeloid, and monocytic leukemia. The 8th Revision (adopted in 1968) maintained the distinction between lymphocytic, myelocytic and monocytic leukemias, each of which was subdivided into acute, chronic, and unspecified; acute leukemia no longer was a separate category. In the 9th Revision (adopted in 1978), the system remained essentially the same. Therefore, prior to 1958, a broad geographic sampling of national or international incidence, mortality, or other descriptive epidemiologic data by histopathologic type is not possible. Most studies of a particular histopathologic group of cases diagnosed before 1958 involved a single hospital or a selected group of leukemics drawn from a few hospitals. Population-based data prior to 1958 included several types of leukemia as a collective group; it might not be possible to separate the results by histologic group.

With some types of leukemia, it is difficult to determine the time of onset. ANLL may present rather insidiously with a preleukemia phase. Whether preleukemia is a separate entity from leukemia or an early smoldering phase of the disease is a somewhat controversial issue. Time of onset is important in determining risk factors and latency period between exposure and subsequent leukemia and for survival analysis.

For leukemia patients with chromosome abnormalities, the sequence of events may be difficult to sort out. If the chromosomal abnormality preceded the onset of leukemia, the chromosome derangement may have been due to a particular exposure (i.e., radiation, chemical, drug). However, because it is not feasible to repeatedly screen large populations for chromosomal abnormalities, it is unlikely that the timing of events can be pinpointed.

Many investigators have tried to identify infectious agents that might play a significant role in the etiology of human leukemia. With the possible exception of the recent work of Gallo and his co-workers at the National Cancer Institute and Miyoshi et al. in Japan, the majority of studies have failed to identify human leukemogenic viruses or other infectious agents that cause leukemia in man [1–6]. Yet substantial indirect evidence indicates that infectious agents may be important. A number of well-characterized viruses have been definitively shown to cause leukemia in cows, cats, gibbons, and other species. Several studies of human acute lymphocytic leukemia have shown weak but repeated evidence of clustering of cases—greater than would be found by chance.

Epidemiologists have used different methods in order to obtain evidence for an infectious etiology in human leukemias. One method analyzes a geographic area within a time-circumscribed interval to determine whether there is clustering among leukemia cases. Another one examines leukemia occurrence in populations heavily exposed to cows or cats with leukemia. A significant problem with both of these methods is case definition, i.e., restriction to one histopathologic type of leukemia or inclusion of several types and possibly even lymphoma. While epidemiologic evidence in humans suggests different etiologies for separate histopathologic types, the animal literature suggests that a single virus can cause more than one type of hematopoietic malignancy in a particular species. But other factors may prevent epidemiologic studies from identifying infectious etiologic agents, even if they are present. The latent period between onset of infection and development of leukemia may be long. The infection may never be clinically apparent or may be trivial and transient. A very small percent of infected humans may develop leukemia. Other promoting risk factors may be difficult to identify. The transmission of the infection may be horizontal, from one infected person to another previously uninfected person, or vertical, from parent(s) to children. It may be difficult to isolate a leukemogenic virus (if it exists) from a person with leukemia because oncogenic infectious agents may not have similar characterists as classic infectious agents; virally caused tumor production may occur in the absence of viral replication. Viruses, viral particles, and viral antigens may not be present in virus-induced cancers, as shown by evidence from study of feline leukemia virus. Gallo and his co-workers have recently identified a retrovirus that appears to be highly associated with rare T-cell leukemias and lymphomas occurring mostly in young adult blacks living in the southeastern part of the U.S. and the Caribbean. Miyoshi et al. have similarly isolated a virus from patients with adult T-cell leukemia, endemic in southwestern Japan. The studies of Gallo et al. have failed to identify the virus in children with ALL [1–6].

Several of these methodologic issues are also relevant to understanding the possible associations of other types of factors. The latency period between exposure to other noninfectious agents and development of disease may be very long, measured in years or decades. A particular chemical or drug or even radiation exposure may cause more than one type of leukemia (perhaps in the same individual), or another type of hematopoietic malignancy in addition to the leukemia. Only a very small percent of individuals exposed to an agent may actually develop leukemia. As yet, no special characteristics identify these "susceptible" subgroups. The exposure may cause no other abnormalities or perhaps only a transient, trivial lowering of blood counts or another manifestation which may be missed if physical or laboratory examination is not performed at the time that a subtle physical or laboratory manifestation of the exposure occurs. Other promoting or interacting factors may be difficult to identify. Finally, an investigator may interpret more than one case of leukemia in a family (which may be due to a particular exposure) as solely a result of "genetic" factors. The genetic substrate may have created a group of "susceptible" individuals whose leukemia was actually caused by an interaction of genetic and environmental factors. Or genetic characteristics may play no role and the leukemias may be a result of an exposure of several of the family members to a particular agent.

Family studies of leukemia may deepen understanding of the role of genetic factors and the interaction of genetic and environmental factors in families with one or more cases of leukemia and may also provide clues about protective factors for leukemia, such as particular HLA types; immunologic profiles; ingestion of particular foods, vitamins, and other nutrients; other lifestyle factors.

Studies of leukemia risk following radiation exposure exemplify the methodological problem of extrapolating data from high-dose exposures to low-dose exposures. Radiation has unequivocally been shown to be leukemogenic at high doses. However, results from studies of low-dose occupational, *in utero,* and fallout exposure to radiation are more controversial. Given the very low incidence of leukemias, sample sizes in many of the studies may not have been adequate. High risk subgroups of particularly susceptible individuals within populations exposed to low-level radiation have not been identified.

As with other types of follow-up studies, appropriate control populations must be selected and bias in follow-up of both exposed and nonexposed populations must be avoided. This may be difficult to achieve because of fixed small sample sizes of exposed populations, control group definitional problems and problems with varying sensitivities of cases and controls to health effects and minor symptoms. As increasing numbers of studies examining associations between chemicals, drugs, other exposures, and the leukemias are carried out, problems similar to those encountered in studies of leukemogenesis following radiation exposure will surface.

The interpretation of effects from several different concomitant exposures is difficult, particularly if risk factors are from more than one exposure category. Gibson et al. noted that the relative risk for leukemia in children 1–4 years old is low (RR = 1.07–1.33) if there was only one risk factor such as history of childhood virus or history of maternal reproductive wastage [7]. For two risk factors from a single exposure category (irradiation of mother before conception and *in utero* irradiation), the risk increased slightly (RR = 1.56). The combinations of multiple risk factors from different exposure categories including reproductive wastage plus childhood viral disease plus *in utero* as well as preconception maternal irradiation substantially increased the relative risk (RR = 4.64) [7]. Perhaps susceptible subgroups consist of individuals who have received a number of key exposures in addition to the risk factor under scrutiny in a particular study. Studies should examine the interaction of several exposures in the etiology of the leukemias.

The epidemiology of the leukemias has received substantial attention compared with other cancers and many nonmalignant diseases. Yet in recent studies, many methodologic and conceptual problems have been encountered. Several studies reported in the 1970's, particularly a few that examined associations between radiation exposure and leukemia, have been widely debated in the epidemiologic and general literature both in terms of results and methodologies employed. As methodologic approaches improve, it is important to review past work, particularly in areas that received extensive attention and study. New, exciting developments in virology, immunology, cytopathology, genetics, toxicology, epidemiologic methodology and statistical analysis should stimulate further etiologic research focusing on the leukemias. Recently Szklo posed the question: Are further epidemiologic studies of leukemia needed [8]? Clearly, the answer is "yes." In addition to overcoming the methodologic problems he described, closer collaboration with experts in virology, immunology, toxicology, etc., should lead to an understanding of risk factors, host factors, and their interaction in the development of leukemia.

REFERENCES

1. R.C. Gallo and F. Wong-Staal, "Retroviruses as etiologic agents of some animal and human leukemias and lymphomas and as tools for elucidating the molecular mechanism of leukemogenesis," *Blood* **60:**545–557 (1982).
2. W.A. Blattner, V.S. Kalyanaraman, M. Robert-Guroff, T.A. Lister, D.A.G. Galton, P.S. Sarin, M.H. Crawford, D. Catovsky, M. Greaves and R.C. Gallo, "The human type C retrovirus, HTLV, in Blacks from the Caribbean, and relationship to adult T-cell leukemia/lymphoma," *Int. J. Cancer* **30:**257–264 (1982).
3. Y. Hinuma, K. Nagata, M. Hanaoka, M. Nakai, T. Matsumoto, K.-I. Kinoshita, S. Shirakawa and I. Miyoshi, "Adult T-cell leukemia; antigen in an ATL cell line and detection of antibodies to the antigen in human sera," *Proc. Natl. Acad. Sci. USA* **78:**6476–6480 (1981).
4. I. Miyoshi, I. Kubonishi, S. Yoshimoto, T. Akagi, Y. Ohtsuki, Y. Shiraishi, K. Nagata and Y. Hinuma, "Type C virus particles in a cord T-cell line derived by co-cultivating normal human cord leukocytes and human leukemia T-cells," *Nature* **294:**770–771 (1981).
5. T. Uchiyama, J. Yodoi, K. Sagawa, K. Takatsuki and H. Uchino, "Adult T-cell leukemia: clinical and hematologic features of 16 cases, *Blood* **50:**481–492 (1977).

6. T. Watanabe, M. Seiki and M. Yoshida, "Retrovirus terminology," *Science* (letter):1178 (1983).
7. R.W. Gibson, I.D.J. Bross, S. Graham, A.M. Lilienfeld, L.M. Schuman, M.L. Levin and J.E. Dowd, "Leukemia in children exposed to multiple risk factors," *N. Engl. J. Med.* **279:**906–909 (1968).
8. M. Szklo, "Are further epidemiologic studies of leukemia needed?" *Am. J. Epidemiol.* **112:**225–231 (1980).

3 Mortality and Morbidity

Etiologic clues may be obtained by examining differences in leukemia mortality by histologic type between countries, between regions within a country, and over time. But mortality statistics by histopathologic type must be examined carefully for errors particularly because of the rarity of each of the leukemias. A common problem with mortality statistics is the incomplete ascertainment of specific causes of death. Significant changes in the International Classification of Diseases (ICD) Coding Systems between 1947 and 1977 and in the interpretation of morphologic, laboratory, and clinical findings may have significantly affected classification of causes of death and may distort apparent secular trends. Until the last 15–20 years, adequate details of distributions of leukemia deaths by histologic type have not been available for studies of international or regional mortality or incidence rates because information was collected on "leukemia" as a single entity [1–3].

Misclassification of different diseases further complicates the interpretation of the mortality and morbidity statistics. Other conditions such as lymphomas, aplastic anemia, multiple myeloma, other undifferentiated cancers, and even leukemoid reactions may be misclassified as leukemias. Comparison of death-certificate diagnoses with corresponding clinical records would reduce the frequency of misclassification, but this is not always feasible. In addition, variation between different populations may be due to differences in the quality of medical care, varying reliability of diagnoses, differences in cultural factors that may affect health behavior, differential frequency in reporting of deaths or cases by officials even after a diagnosis has been made.

Real differences in rates may be due to variation in genetic susceptibility, in environmental factors, or in a combination (or interaction) of host and environmental characteristics. Examination of descriptive statistics may suggest hypotheses about potential associations or etiologic factors that deserve additional investigation. Nations or regions with high mortality or morbidity rates may contain population subgroups at high risk for one or more of the leukemias, a situation which should be studied in more detail.

In general, the variation among different countries for the leukemias is less than for most types of cancer [4, 5]. Differences between countries with high and low rates are rarely greater than twofold. Analysis of these rates by cell type does reveal some differences that are not understood. Speculation about the reasons for such differences has led to hypotheses that deserve additional investigation by histopathologic type-specific studies.

MORTALITY

International

ALL Court-Brown and Doll analyzed leukemia (all types combined) mortality rates among children and young adults under the age of 30 in England and Wales between 1911 and 1959. Data provided by the Registrar General of England and Wales enabled the deaths occurring from 1945 through 1959 to be classified by histopathologic and clinical subtype; breakdown by type was not available in previous years [6]. The majority of leukemia deaths occurring after 1945 were ALL. Age-specific mortality rates peaked at age 3, declined as age increased except for a small peak at ages 15–19, and reached their lowest levels at ages 25–29 before beginning to rise slowly in adulthood. Although the age-specific mortality pattern did not substantially change over the years, between 1911 and 1959 the mortality rates did increase steadily in each 5-year age group and for both sexes. While part of the overall increase may have been artifactual as a result of improved diagnoses, part of the increase was probably real. In England prior to 1910, and in the U.S. during the 1920's and 1930's [7], there was no evidence of an early childhood peak in mortality. Though the published U.S. data is not cell-type specific, the majority of childhood acute leukemias in the U.S. are ALL and should be comparable to data from England and Wales. The childhood peak in the U.S. first became apparent between 1921 and 1939 and was more pronounced after 1940. Using the U.S. data, Slocumb and MacMahon [8] analyzed this childhood peak at age 3 and noted that it was actually the result of a decline in mortality up to age 2. The decline began during the 1920's and 1930's among infants under 1 year. In subsequent years, the rates also declined in children over age 5. The changing age pattern was probably due to increased duration of disease and postponement of death as a result of improvements in therapy. The mortality rate for ALL was similar for both sexes under age 1; in older age groups, male mortality was higher. Examination of time trends showed that the early childhood peak has existed in England and Wales since 1910 [6].

Mainwaring examined childhood leukemia mortality rates in Liverpool during 1955–64 and found those for ALL were 3 times higher than those for ANLL [9]. The rates peaked at 4–5 years of age; the higher male/female ratio (1.83:1) was in excess of those noted in England and Wales for a similar period. The suggestion of a slight increase in occurrence of ALL with summer onset was not statistically significant. The study noted that cases tended to occur more frequently in certain parts of the city where older housing and overcrowding were more common.

Hems and Stuart analyzed data from death certificates of 978 children who died from leukemia in Scotland during 1939–68 [10]. Their findings were similar to those of Court-Brown and Doll which indicated that leukemia mortality, principally from ALL, increased about threefold for all age groups during the 1940's and 1950's and declined slightly in the 1960's. Mortality from ALL peaked at 3–4 years of age; overall, ALL was 3 times more common than ANLL. Compared to those of other countries, leukemia mortality rates for Scottish children were low. No evidence was found for an urban–rural gradient, seasonal variation, or clustering of cases.

Klaassen reviewed all deaths from childhood acute leukemia in Saskatchewan, 1961–66, ascertained from records in all clinics in the province, pathology reports as well as death certificates [11]. The ratio of ALL to ANLL deaths was 2.7:1. He compared these leukemia rates to those obtained by Meighan [12] during 1947–60 and found an increase for all ages which he believed to be real. The age and sex distribution of the deaths also did not change. A reason for the increased rate was not apparent.

ALL is much less common than ANLL in Japanese children of all ages. The incidence and mortality rates of ALL was only one-third that of ANLL in a study carried out in 1965–66 [13]. In general, lymphocytic leukemias are relatively infrequent in Japan. The decreased frequency of ALL compared to ANLL was also noted in China [14]. Data obtained from Children's Hospital, Soochow, China, indicated that, of 173 children under age 15 with leukemia, 50% had ALL in comparison to 80% in the U.S. [15].

ALL is very rare under 5 years of age in all parts of Africa including west [16–20], east [20–24], central [25], and south [26]. The peak mortality rates observed in ages 3–5 in western countries is absent in those African populations that have been studied. Several explanations have been advanced for these findings. Some investigators believed that ALL is a variant of Burkitt's lymphoma. Others disagree because the incidence of Burkitt's lymphoma varies among different countries in Africa; in some countries, the incidence of Burkitt's is higher than that for ALL in children under age 5, while in others it is lower. The two diseases also differ in morphology and response to treatment. Some investigators have suggested that a high rate of infectious diseases occurring early in childhood may somehow either prevent ALL or delay its onset. Findings of low leukemia mortality rates in many countries with correspondingly high death rates from communicable infectious diseases have been used to support this theory [17, 19, 20].

However, the rarity of diagnosed cases of ALL (and other leukemias) in Africa may also reflect the following factors: (1) the inability of patients to travel great distances to a hospital; (2) insufficient trust in physicians to seek their care; (3) cost of medical care; (4) unwillingness to forfeit a day's work to stand in a line at a clinic; (5) misdiagnosis by physicians who fail to differentiate leukemia symptoms from other more frequently seen infectious diseases; (6) underdiagnosis or incorrect diagnosis by pathologists; and (7) underreporting of cases by physicians to registries. Nevertheless, the relative rarity of ALL in most parts of Africa is believed to be real [16, 18, 24, 27].

CLL Court-Brown and Doll also examined CLL mortality data in England and Wales from 1945–49 through 1955–57 [28, 29]. Above age 30, mortality increased rapidly with age (as much as 600-fold), for both sexes with only a minor decrease occurring in males ages 85 and over compared to males 80–84, but not consistently in females in these two age groups. For both males and females ages 65 and older, CLL mortality increased between 1945–49 and 1955–57 by a factor of 1½–5, depending upon the age group. Of all of the leukemias, the male to female ratio is highest for CLL (1.7–2.0:1) and lowest for CML (1.0–1.1:1). Keogh and McCall examined data from CLL deaths during 1946–55 in Melbourne, Australia, and found a similar increased mortality in older males and females compared to that of earlier time periods [30]. The apparent increase in death rate in the older age group may result from improvements in diagnosis and medical care for the elderly.

Nishyama noted that CLL constitutes less than 2% of leukemia in Japan. Mortality rates in Japanese increase with age beginning at 30–39 years (0.8 per 100,000 per year) with the highest rates occurring in the 70 and over age group (15.3 per 100,000). CLL is virtually confined to adults in Japan and is rare under the age of 40. The low leukemia mortality rate as a whole in Japan may well be due to the very low mortality from CLL, which is 39.3 times higher in the United States, 41.0 times higher in Israel, and 91.8 times higher in Sweden than in Japan [13].

Haenszel and Kurihara compared the mortality rates of Japanese migrants to the U.S. with rates of Japanese, U.S.-born Japanese, and U.S. whites [31]. Despite differences in rates by histologic type, childhood and young adult mortality rates in Japanese migrants for all leukemias combined are quite similar to those of native Japanese. The mortality experience of U.S.-born Japanese descendants is also similar to native Japanese. However, the proportions of the various histologic types do not vary greatly between native Japanese and U.S.-born Japanese descendants. For native Japanese, CLL constitutes 2.9% of all leukemias in persons over 55; CLL in U.S.-born Japanese descendants accounts for 5% of leukemia deaths in the same age group, whereas U.S. whites with CLL constitute 31.5% of leukemia deaths.

In New Zealand, Gunz noted a continuous increase in leukemia mortality for all age groups from 1908 to 1963, with the most significant increase in the 75–79 age group. Examination of the data showed that the increasing mortality probably reflected an increased ascertainment of CLL, CML, and all acute leukemias since all cell types were involved rather than one particular type. Gunz also noted that CLL was rare among Maoris, a Polynesian race constituting approximately 7% of the New Zealand population. As the Maori population has a young age distribution, the significance of this observation is not clear [32].

Reports of several CLL patient series from Zaire, Kenya, Uganda, and Rhodesia indicate age distributions and sex ratios similar to those from western nations [22, 23, 25, 26, 33]. However, another series from Zaire reported that over 60% of the CLL cases were under age 40 at the time of diagnosis [34]. Several series from West Africa noted a large proportion of cases diagnosed in the fifth decade as well as equal frequencies of males and females [16, 19, 35, 36]. In many of the clinical series reported from Africa, splenomegaly is very pronounced and lymphadenopathy is minimal or absent.

In a review of several series from different parts of Africa, Fleming found that splenomegaly in African patients with CLL does appear to be more pronounced than in western CLL cases [20]. He also believes that there may be a significant amount of misdiagnosis of African patients especially when infectious diseases, benign lymphoid conditions, tropical splenomegaly syndrome, and various immunologic disorders have not been properly excluded.

ANLL In England and Wales, Court-Brown and Doll found that childhood and young adult ANLL mortality rates did not change with age as did those of ALL from 1911 through 1959, except for two small peaks at ages 2–4 and 15–29 [6]. However, beginning in the mid-1940's, there was a real increase in ANLL mortality for all ages.

In England and Wales, Scotland, and Canada, ANLL mortality rates are very low in childhood and increase progressively with age with the rise beginning around age 35 [6, 10, 11, 37]. A small peak is evident in males in their late teens. In some studies, a smaller peak was noted for females, but this was not a consistent finding. Court-Brown and Doll found a 33% increase in mortality from ANLL for adults of all ages and both sexes between 1945–49 and 1955–57 [28]. This was also noted by Klaassen for the period 1947–60 and 1961–66 [11].

Hems and Stuart noted that ANLL mortality rates for females ages 0–14 years old in Scotland increased by 2½ times between the 1940's and the 1960's, but rates for males remained constant [10]. In contrast, the rates increased twofold for both males and females in this age group during a similar time period in England and Wales [6, 28]. The disparity by sex in the Scottish series is difficult to explain.

In Japan, myelocytic leukemias, both acute and chronic, have incidence and mortality rates 2–20 times higher for all age groups compared to rates for acute and chronic lymphocytic leukemias [13, 38]. Nishiyama showed that the ANLL rate in those under age five in 1965–66 was 3 times higher than that of ALL.

The pattern of age-specific mortality for ANLL appears to differ markedly in several parts of Africa compared to that in Europe and North America. Mortality rates are highest in the second decade of life in southern Africa [39] and in the third decade in east Africa [22, 23, 33] though it is also not infrequent in childhood. The predominance of ANLL in children and young adults is also found in many other parts of Africa.

With data from Johannesburg, South Africa, Kusman et al. confirmed the difference in ratio of ANLL to ALL (1.0–1.5:1) in Africa compared to that seen in western countries (1:3) [40]. They found that 50% of their cases of acute leukemia in childhood were myelomonocytic. Similar frequencies were also reported by Jeffrey et al. in Rhodesia [26] and by Davies et al. in Uganda [27].

Chloroma, a mass of ANLL cells, occasionally develops during the late stages of ANLL. It occurs with great frequency in north [41, 42], west [18, 19], and central Africa [25] but not in southern Africa [26, 39]. Since this condition was once more frequent in developed countries, its presence in Africa may represent a late development in the natural history of the disease and reflect a lack of treatment early in its course.

CML Court-Brown and Doll examined CML mortality rates in England and Wales for three time periods: 1945–49, 1950–54, and 1955–57 [28, 29]. For all time periods, age-specific rates steadily increased to a peak at ages 60–75 and then decreased in those 80 and over. Between 1945–49 and 1955–57, male mortality from CML increased in the age groups 35 and above. However, at age 60 and above, the rates nearly doubled for males, with the increase in females being confined to those 60 and older.

Nishiyama examined the frequencies of various types of leukemia in Japan by studying 3,925 autopsied cases [13]. Approximately 15% of the leukemia cases were chronic myelogenous leukemia, compared to 20% in the United States. CML mortality rates increased with age, from 4.6 per 100,000 per year for those ages 0–9, to 7.45 per 100,000 for those 60–69. For those over 70, the rate declined. Mortality is higher for males than for females. However, mortality rates for CML in Japan are low compared to those for the United States, which has a total mortality rate 2.9 times higher, Israel a rate 2.2 times higher, and Sweden a rate 5.3 times higher than that of Japan.

Tomonaga indicated that mortality from all leukemias in Japan was about one-half of that for North American and European countries [43]. When age-specific CML mortality rate curves of atomic bomb survivors from Hiroshima and Nagasaki were compared with those from other parts of Japan, the age-specific mortality of CML in the atom-bomb exposed group was skewed to younger ages, especially in Nagasaki.

Very little data separately listing either age-specific or age-standardized rates for CML in African countries have been reported. Only two African counties were included in *Cancer Incidence in Five Continents*—Nigeria, with data reported from the cancer registry in Ibadan, and Rhodesia, represented by cancer cases collected by the Bulawayo Cancer Registry operating from one large regional hospital offering specialist services to the African population of western Rhodesia [44]. For several reasons, including political and social instability and strong rural ties of the indigenous African population, substantial use of witchcraft in lieu of modern medical treatment, and very low utilization of hospital medical resources by older women, there is probably significant underregistration in the Bulawayo Registry.

Comparison of age-standardized rates for CML in Ibadan, Nigeria (males 1.0 per 100,000; females 0.6) and Bulawayo, Rhodesia (males 0.1 based on less than 10 cases; females 0.2) with those for whites and blacks of a U.S. city (Detroit) included in the Five Continent survey (white males 1.8; black males 1.2; white females 0.8; black females 1.2) show comparable rates for Nigerian males and U.S. blacks, with a somewhat higher rate in U.S. white males and much lower rate (probably due to under ascertainment of cases) in Rhodesian African males. Nigerian female and Detroit white female rates were comparable although the Detroit black female rate was twice that among Nigerians; similar to the situation for males, the Rhodesian African female rate was substantially lower than all of the other female rates [39]. CML rates could not be found for other African countries.

Edington and Hendrickse reported that the frequency distribution for reticuloendothelial tumors reported to the Ibadan Registry showed that CML constituted 9.4% of these, CLL 7.7%, ANLL 6.0%, and ALL 1.3% [17].

While clinically and morphologically CML does not differ in Africa from Europe and the United States, it usually appears at a younger age in Africans. Although age-specific rates are not available, clinical series suggest that peak occurrence appears to be in the third and fourth decades compared to the fifth decade in the United States [16, 19, 20, 22, 24–26, 33]. This difference may reflect differences in the age structure of most African populations compared with that of the U.S.

In Africa, there is a male preponderance by a ratio of 2:1. However, the ratios vary from a high of 6:1 in Zaire [25] to 3:1 in Rhodesia [26]. Other authors have found an equal sex ratio in Uganda [23] and northern Nigeria [19]. Lowenthal and Essien pointed out that male predominance was more marked in younger age groups [35, 45].

Summary International mortality data appear to show that crude death rates for all types, and particularly ALL and ANLL, increased steadily for all age groups beginning in the early 1900's and then began to level off in the 1950's in some countries and in the 1960's in others. Age-specific mortality rates for ALL in most countries showed a peak in childhood that shifted from ages 1–2 to ages 3–4 and even older after 1950. ANLL has shown no such consistent childhood peak. Male death rates have always been greater than female death rates for most leukemias with the possible exception of CML, but always the male/female ratio was less than 2:1. The incidence rates of ALL and CLL are low in Japan, China, and other Asian countries. ALL is also rare in nearly all African nations. Overall, leukemias are infrequent in Africa but there are many more lymphomas. In combined European series, the ratio of leukemias to lymphomas is 3:1, whereas in combined African series, the ratio is 1:10 [44]. Some of this discrepancy may represent underdiagnosis of leukemia in Africa, particularly for ALL in children, but most investigators believe this to be unlikely. Two possible explanations for the low rates of leukemia are (1) a lower frequency or even absence of known risk factors (i.e., excessive maternal age, exposure to xrays, Down's syndrome, minor congenital abnormalities) or (2) an increase in resistance of the African child to leukemia.

United States

Examination of regional differences in leukemia mortality and incidence within countries or between different geographic areas may provide clues to the etiology of the leukemias. Several mortality studies of the geographic distribution of leukemia in the United States at different time periods have examined all leukemias combined, rather than separately by histopathologic group [46–54]. Histopathologic type-specific population-based mortality data may also be limited because classification schemes for both clinical and death certificate coding have changed repeatedly over the years and many death certificates list "leukemia" as the underlying cause of death without specifying histopathologic type. Therefore, even histopathological type-specific death statistics may be limited, even in national studies. The basic trends for the leukemias as a group will be briefly summarized, then histopathologic type-specific mortality data will be presented.

Leukemia mortality in 1949–53 was higher in the northern, central, and western states than in the south and southeastern states [46]. This geographic gradient was observed for both whites and nonwhites, males and females. Most of the excessive leukemia mortality in the "high mortality" regions represented increased mortality in New York State, Minnesota, and California. MacMahon found that this geographic pattern was less marked in 1949–53 than in 1938–42, and that the geographic gradients were highly correlated with density of physicians, which may imply that differences in medical care and diagnosis may be responsible for the increase rather than indicate true differences in mortality [47, 48].

The 1950–69 leukemia mortality data from the *Atlas of Cancer Mortality by County* indicate the highest mortality rates occur in a central band extending from Minnesota through Texas for white males and females; and to a lesser extent, for nonwhites [50, 51]. Nonwhites tended to have higher mortality rates in the northeast than in other areas. The lowest mortality rates for both sexes were in the southeast and southwest. If this patttern were applicable to all histopathologic types of leukemia then one might attribute the differences to medical care and diagnosis. However, incidence data from the Third Cancer Survey suggests otherwise, as described below in the section on incidence in the U.S. [55].

Blair et al. examined U.S. geographic variation by county for leukemia, all types combined, and separately by histopathologic type. For whites, high-rate counties were characterized as follows: urban, higher socioeconomic levels, a larger proportion of the population was foreign-born, and more workers were employed in the food-processing, printing, fabricated metal product and machinery industries. Among males, 30/57 of the high-rate counties were located in north and south central regions of the country; among females, 17/37 high-rate counties were in the north and south central regions. Examination of geographic patterns by histopathologic type, available only for the four-year period 1968–71, showed higher rates of AML in the north and south central regions. Patterns for the other leukemia subtypes showed no major variation by region [54].

These investigators noted other histopathologic-type differences. For

ALL, high-rate counties for males had a smaller rural farm population and fewer persons of German, Scandinavian, and Russian ethnicity than other U.S. counties. High-rate counties for females had less involvement in food manufacturing and corn production. For CLL, high-rate counties for males had a higher median income and heavier corn production and herbicide use. Among females, these counties had a lower median income and lower corn production and herbicide use. For AML, high-rate counties for both males and females had a more urban population, higher median income, a larger proportion of the population of German and Scandinavian origin, and a heavier concentration of the food industry. For males only, the high-rate AML counties had a higher proportion of acres treated with herbicides. For CML, high-rate areas among males had a more urban population, higher median income, and a smaller population of German and Scandinavian origin whereas among females they had a higher median income, a larger population of German origin, and more acres planted with corn treated with herbicides [54].

Burbank examined age-, sex-, and race-specific mortality rates in 1950–67 for two major groupings of the leukemias (lymphatic and myeloid) as determined by the ICD 7th Revision. No further distinction was made for these two groups between acute and chronic types. For lymphatic leukemia, he noted a small peak in mortality in the 5–9 year age group; rates then fall until ages 45–59 after which they rapidly increase. Rates were higher for males than females at all ages. Around age 70, rates began to decline for nonwhites but continued to increase for whites. Over the interval 1950–67, mortality rates for whites ages 0–9 declined significantly, presumably due to improvement in treatment and response in whites with ALL, but not in nonwhites. Mortality rates did not change over the interval 1959–67, when the coding was revised (6th to 7th). Geographic trends for lymphatic leukemia were similar to those for all leukemias: decreased death rates in the south and northeast and increased rates in the central and north midwest. For myeloid leukemia, death rates rose continuously with age for all four sex–race categories until ages 60–64 when rates declined for female nonwhites and plateaued for male nonwhites but continued to rise for whites. From 1959 through 1967, age-adjusted death rates rose significantly for male nonwhites and for female whites. The only geographic variation noted was a decrease in death rates for white males in the central midwest and the west [52].

Stark and Oleinick analyzed leukemia mortality rates separately by histopathologic type, age, sex, race, year of death, and urban–rural location for 20,711 leukemia deaths in the U.S., 1950–59. Age-specific and age-adjusted rates were higher for white males compared with white females in all histopathologic-type groups. In whites ages 0–20, lymphocytic leukemia mortality was higher than myelocytic leukemia. The greatest difference was at ages 1–4; whites had mortality rates from lymphocytic leukemia that were 2–8 times greater than from myelocytic leukemia. For all whites, age-adjusted mortality was higher in urban areas than in rural areas. Age-adjusted rates for white urban children increased between 1950–53 and 1954–57, while rural rates decreased. With these opposite trends, total rates did not change over this interval. For nonwhites, male rates were substantially higher than female rates in every age group except those under age 1. For lymphocytic leukemia, age-specific rates for all ages were considerably lower in nonwhites than in whites. No age peak was noted for nonwhites with lymphocytic leukemia in the age group 1–4. There was much less disparity between whites and nonwhites for myelocytic leukemia [53].

More recently, Selvin et al. analyzed histopathologic type-specific leukemia mortality for the U.S. white population using death certificates collected by the National Center for Health Statistics for 1969 through 1977 and coded by the 8th Revision of the ICDA. Estimates of race-, age-, and sex-specific county intercensal population data made by the U.S. Census Bureau for 1970–77 were used as denominators in the calculation of rates. These investigators noted that, for ALL, childhood mortality rates rose to a peak in the 5–9 year age group, then declined with an antimode between 35–44 years, then increased geometrically with age; trends were similar for males and females but rates were consistently lower at all ages for females. (Since leukemia type-specific mortality rates, 1969–77, are very similar to incidence rates from the Third National Cancer Survey, 1969–71, graphs depicting mortality are not presented; the reader is referred to Selvin et al. [56]. Figures and a table of incidence rates are presented on pages 38–40.) Over the nine-year period, substantial decreases in ALL were noted for almost all age and sex subgroups with the exception of males ages 10–14. This was considered consistent with the increasing efficacy of chemotherapy over this time period. Geographic patterns were not found to be very striking, although high rates were noted for adults in the middle regions of the country and low rates on the east coast. Similar geographic distributions were found for both ALL and AML in children but not adults. A substantial urban excess was noted for ALL in adults as was a positive association with altitude. CLL mortality in males rose geometrically from very low rates in the age group 35–44 to very high rates in the oldest age group, 85 and over. Although the pattern was similar for females, the male to female sex ratio ranged between 2.0 and 2.6. Rates did not change significantly over the nine-year time period. High rates for CLL were found in the north central area and low rates throughout the south, most noticeably in adults over age 65. For both males and females with AML, peak childhood rates were found in the age group 0–4, dropped slightly to a low in ages 5–9, then rose almost geometrically until ages 75–84 after which the rise was less steep. Rates for AML increased over the time period 1969–77 for both males and females in all but the youngest age groups. High rates for AML in adults of both sexes were found on the west coast. For both childhood and adult AML, the urban-rural ratios were less than one. Age-specific rates for CML rose geometrically from very low rates in the age group 15–24 to high rates in the oldest age groups for both males and females. For males with CML, rates increased 32% in the 45–54 year age group and to a smaller extent in older age groups; female rates decreased slightly among all adult age groups. There was no distinct geographic pattern. For all histopathologic types, there was a lack of geographic association between childhood and adult forms of leukemia. Selvin et al. postulate that the geographic correlation between childhood ALL and AML may reflect a similar etiology, whereas the geographic variability of the four histopathologic types in adults suggests that different risk factors are responsible for each type.

INCIDENCE

International

There are several problems in comparing the incidence of leukemia (as well as other cancers) in different populations throughout the world. Differences in methods by which cases are ascertained and reported (i.e., physician reports, pathology reports, hospital discharge listings, autopsy records, hospital and regional tumor registry listings, death certificates, etc.) may result in varying levels of ascertainment between countries. Within a country, the particular method(s) of ascertainment used may result in varying estimates of the true proportions of the different leukemia cell types. The cell type of the leukemia is frequently given in tumor registries. But a substantial proportion of death certificates lack this information and are coded as "other and unspecified leukemia." If the proportion of cases that are registered from death certificates without specification of histopathologic type is high, then the recorded incidence rates for each of the cell types are likely to be artificially low. When patients are both diagnosed and followed up solely as outpatients, as frequently occurs with CLL, cases may not be found in hospital discharge listings, pathology reports, and autopsy records and are therefore less likely to be reported to registries [44].

These problems are present in nearly all countries, but are especially serious with data obtained in several countries in Africa. Information about leukemia incidence rates is spotty, problematic as to numerator and denominator because of incomplete or missing vital statistics, and lacking cell-type specificity. Older Africans are more likely to falsify their age and use traditional "medicine men." They may not come to the attention of modern medical practitioners nor be reported to cancer registries, and therefore the cause of death on their death certificates may be incorrect for one or more of these reasons [20, 44]. A cancer survey done in Ibadan in Western Nigeria agreed with other findings that the incidence of cancer as a whole in the older African is much less than that found in the United States [17]. But the inaccuracy of age reporting in the older members of African society, the younger age distribution in Africa than in the U.S. and Europe [44], and the lack of cell-specificity in the cancer rate surveys in Ibadan and other regions in Africa make data interpretation difficult.

Despite these methodologic problems, in 1966, the International Agency for Research on Cancer (IARC) published cancer incidence rates representative of many geographic areas and countries of five continents, then updated the incidence rates in 1970 and 1976. Inclusion of data is dependent upon: reliability of both numerator and denominator information, adequacy and completeness of reporting of cancer cases (as judged by tests of internal consistency and stability of data), the proportion of histologic and death certificate diagnoses, correlation between calculated incidence rates with available mortality data, and availability of a recent reliable census of the population-at-risk [44].

The most recent cancer incidence rates in Volume III of the IARC publication have been based on cancer data obtained in the various geographical areas and countries at some period between 1960 and 1973. Cancers were coded according to the ICD 7th and 8th revisions.

ALL In general, there is minor international variation in ALL incidence rates with a few exceptions. Higher incidence of ALL in males of all age groups compared with females occurs in most countries with the exception of the only South American countries in the survey, Brazil and Columbia. In these two countries and in Spanish-speaking populations of New Mexico and El Paso, Texas, a slight female predominance is noted for ALL subgroups of all ages [44] (see Table 3-1). Within-country variation is also minimal with two exceptions (as seen in Table 3-1). Between two Canadian provinces, a tenfold difference in age-standardized rates in males was seen; annual incidence of ALL is 2.5 per 100,000 in Manitoba and 0.2 per 100,000 in Newfoundland. Rates are 10 times higher for female Jews (1.1) living in Israel compared with female non-Jews (0.1); the differences are not as marked for males.

CLL Incidence rates for CLL are fairly uniform for whites in western Europe and North America, ranging from 0.9 to 2.4 per 100,000 per year (Table 3-2). Lower rates occur in Spain, Brazil, Columbia, Puerto Rico, and persons of Spanish origin in New Mexico and range from 0.2 to 0.4

Table 3-1 Acute lymphocytic leukemia (age-standardized incidence rates by country for males and females)

	M	F		M	F
Nigeria, Ibadan	0.3	0.1	Hungary		
Bulawayo, Africa	3.2	0.2	Vas	1.1	0.4
Brazil, Sao Paulo	0.4	0.6	Szabolis	0.7	0.6
Canada			Malta	1.8	1.1
Alberta	1.7	1.2	Norway		
British Columbia	0.9	0.8	Rural	0.7	0.3
Manitoba	2.5	1.5	Urban	0.6	0.2
Maritime Provinces	1.5	1.3	Poland		
Newfoundland	0.2	0.2	Katowice	0.3	0.2
Quebec	0.8	0.6	Warsaw, City	0.1	0.2
Saskatchewan	1.8	1.8	Warsaw, Rural	0.2	0.0
Colombia, Cali	0.3	0.8	Spain, Zaragoza	0.5	0.3
Puerto Rico	1.3	1.1	Sweden	0.7	0.5
India, Bombay	0.6	0.3	Switzerland, Geneva	1.5	1.9
Israel			United Kingdom		
All Jews	1.0	1.1	Oxford	1.4	1.2
Born Israel	0.9	0.8	Sheffield	1.0	1.0
Born Europe or America	0.7	0.7	SMCR 1967–71	1.1	0.7
Born Africa or Asia	0.5	1.0	Ayrshire	2.1	0.9
Non-Jews	0.7	0.1	Birmingham	1.4	1.0
Japan			Yugoslavia, Slovenia	0.4	0.4
Osaka	0.7	0.7	New Zealand		
Miyazi	0.4	0.3	Maori	1.2	2.0
German Democratic Republic	0.0	0.0	Non-Maori	1.5	1.1
Federal Republic of Germany, Hamburg	0.3	0.1			

Source: Cancer Incidence in Five Continents [44].

per 100,000 per year. Other countries with low rates are Poland and Hungary in Eastern Europe (0.2), India (0.2), and Israel (where Israeli-born Jews and non-Jews have a low incidence (0.4) but Jews born in Europe, America, Africa or Asia have an age-standardized CLL incidence of 1.1–1.3 per 100,000 per year). The Japanese also have a very low incidence of CLL compared with western populations. CLL represents only 2.5% of all leukemias seen in Japan compared with 30% in western countries [13]. Genetic factors or maintenance of microenvironmental factors across generations may be responsible for the low rate of CLL in Japan, since Japanese migrants to the U.S. continue to have very low incidence rates in succeeding generations [17].

The low frequency of CLL in Japan may result from the infrequency of the disease in Japanese who are 55 years or older [13, 57–59]. Within Japan, rates vary, increasing from north to south. The highest incidence in Japan (Nagasaki) is greater than 2.5 times that in the lowest incidence regions. A difference in the genetic make-up of the population of Nagasaki may be partially responsible for the difference in incidence rates. Nagasaki

Table 3-2 Chronic lymphocytic leukemia (age-standardized incidence rates by country for males and females)

	M	F		M	F
Nigeria, Ibadan	1.4	2.1	Hungary		
Bulawayo, Africa	4.3	1.3	Vas	3.6	0.9
Brazil, Sao Paulo	0.4	0.3	Szabolis	0.3	0.2
Canada			Malta	0.0	0.1
Alberta	3.4	1.3	Norway		
British Columbia	2.1	1.1	Rural	1.0	0.5
Manitoba	2.8	1.4	Urban	1.0	0.5
Maritime Provinces	1.3	0.5	Poland		
Newfoundland	0.5	0.4	Katowice	0.5	0.2
Quebec	1.8	0.9	Warsaw, City	1.2	0.5
Saskatchewan	4.5	1.9	Warsaw, Rural	0.8	0.2
Colombia, Cali	0.3	0.4	Spain, Zaragoza	0.2	0.2
Puerto Rico	0.5	0.4	Sweden	2.7	1.3
India, Bombay	0.4	0.2	Switzerland, Geneva	2.5	0.8
Israel			United Kingdom		
All Jews	2.2	1.2	Oxford	1.5	0.7
Born Israel	2.1	0.4	Sheffield	1.3	0.6
Born Europe or America	2.4	1.3	SMCR 1967–71	1.4	0.7
Born Africa or Asia	1.2	1.1	Ayrshire	0.8	1.0
Non-Jews	1.1	0.4	Birmingham	1.4	0.7
Japan			Yugoslavia, Slovenia	2.0	1.2
Osaka	0.1	0.1	New Zealand		
Miyazi	0.1	0.0	Maori	1.1	0.0
German Democratic Republic	2.2	1.0	Non-Maori	2.1	1.0
Federal Republic of Germany, Hamburg	1.0	0.6			

Source: Cancer Incidence in Five Continents [44].

was the only port open to foreign trade during Japan's prolonged period of isolation (1635–1859). Data from China confirm a low CLL incidence rate in Asians: only 5% of all leukemias are CLL compared to 30% in the United States [14]. In Israel, incidence rates are lower in non-Jews (males 1.1, females 0.4) than Jews (males 2.2, females 1.2). For Israeli Jews, rates were lower for males born in Africa or Asia (1.2) than those born in Israel (2.1), Europe (2.4), or America (2.4). The difference in rate by birthplace does not hold true for Israeli females [44]. In a study carried out at the South African Institute for Medical Research (responsible for most histopathologic diagnoses in the Transvaal), all specimens with the diagnosis of a lymphoproliferative neoplasm were examined and classified using the criteria and nomenclature of Rappaport [59]. With this data, age-standardized rates per 100,000 per annum were constructed for various lymphoproliferative neoplasms to compare well-differentiated lymphoproliferative malignancy rates in whites versus Bantus. These specimens included cases of well-differentiated lymphocytic lymphoma (WDLL) and chronic lymphocytic leukemia because physicians and pathologists submitting the

Table 3-3 Acute myelocytic leukemia (age-standardized incidence rates by country for males and females)

	M	F		M	F
Nigeria, Ibadan	0.4	0.3	Hungary		
Bulawayo, Africa	0.9	1.0	Vas	0.2	0.0
Brazil, Sao Paulo	0.7	0.5	Szabolis	0.5	1.0
Canada			Malta	0.5	0.5
Alberta	1.3	1.1	Norway		
British Columbia	0.8	0.7	Rural	1.4	1.7
Manitoba	1.9	1.5	Urban	1.9	2.0
Maritime Provinces	1.7	1.1	Poland		
Newfoundland	1.3	0.6	Katowice	1.3	1.1
Quebec	1.2	0.8	Warsaw, City	1.3	1.8
Saskatchewan	1.6	1.2	Warsaw, Rural	1.1	1.1
Colombia, Cali	0.8	0.6	Spain, Zaragoza	0.2	0.0
Puerto Rico	1.7	1.3	Sweden	1.2	1.4
India, Bombay	0.4	0.3	Switzerland, Geneva	2.7	3.7
Israel			United Kingdom		
All Jews	1.7	1.3	Oxford	1.7	1.9
Born Israel	2.3	1.5	Sheffield	1.2	1.4
Born Europe or America	3.2	1.2	SMCR 1967–71	1.2	1.5
Born Africa or Asia	3.1	0.7	Ayrshire	1.8	1.1
Non-Jews	1.8	0.4	Birmingham	1.9	1.9
Japan			Yugoslavia, Slovenia	0.9	0.9
Osaka	1.3	1.4	New Zealand		
Miyazi	1.7	2.1	Maori	2.5	0.9
German Democratic Republic	2.6	2.4	Non-Maori	2.4	1.5
Federal Republic of Germany, Hamburg	1.0	1.0			

Source: Cancer Incidence in Five Continents [44].

material often did not note whether the individual clinically had CLL or WDLL. For both males and females, age-standardized rates were higher in whites (males 0.67 per 100,000; females 0.38) than Bantus (males 0.45; females 0.16). But frequency of medical certification of cause of death in whites was estimated to be 99% whereas it was much lower in Bantus, particularly for those living in rural areas.

ANLL Higher male than female age-standardized rates for ANLL have also been found in virtually all countries included in the Five Continent Incidence Study (Table 3-3). Somewhat higher ANLL rates are noted in more developed countries and lower rates are reported in underdeveloped nations. This trend is not seen with the other cell types of leukemia and may represent a real association between ANLL and industrialization, diet, socioeconomic factors, or other characteristics of developed countries that distinguish them from underdeveloped countries [44].

CML CML age-standardized incidence rates are higher in males than in females for most countries except East Germany, Hungary, and Swit-

Table 3-4 Chronic myelocytic leukemia (age-standardized incidence rates by country for males and females)

	M	F		M	F
Nigeria, Ibadan	1.0	0.6	Hungary		
Bulawayo, Africa	0.1	0.2	Vas	0.6	0.8
Brazil, Sao Paulo	0.5	0.5	Szabolis	0.5	0.6
Canada			Malta	0.3	0.6
Alberta	1.0	0.5	Norway		
British Columbia	0.6	0.5	Rural	0.7	0.5
Manitoba	1.2	1.1	Urban	0.7	0.4
Maritime Provinces	0.7	0.4	Poland		
Newfoundland	0.3	0.2	Katowice	0.3	0.2
Quebec	0.7	0.6	Warsaw, City	0.7	0.4
Saskatchewan	1.3	0.8	Warsaw, Rural	0.3	0.2
Colombia, Cali	1.2	0.6	Spain, Zaragoza	0.4	0.0
Puerto Rico	1.0	0.8	Sweden	1.3	1.0
India, Bombay	0.6	0.7	Switzerland, Geneva	0.7	1.0
Israel			United Kingdom		
All Jews	0.7	0.9	Oxford	0.9	0.6
Born Israel	0.7	0.6	Sheffield	0.6	0.6
Born Europe or America	0.7	1.0	SMCR 1967–71	0.7	0.5
Born Africa or Asia	0.6	0.6	Ayrshire	0.5	0.6
Non-Jews	0.2	0.7	Birmingham	0.7	0.7
Japan			Yugoslavia, Slovenia	1.1	0.8
Osaka	0.5	0.2	New Zealand		
Miyazi	0.3	0.3	Maori	1.0	0.3
German Democratic Republic	1.5	1.2	Non-Maori	1.0	0.6
Federal Republic of Germany, Hamburg	0.6	1.0			

Source: Cancer Incidence in Five Continents [44].

zerland, where the reverse is true [44]. Otherwise, there is little variation in the rates between countries for either males or females (see Table 3-4).

United States

U.S. population-based cancer incidence rates separately determined by histopathologic type of neoplasm have only been available since the mid-1970's following analysis of data from the Third National Cancer Survey (1969–71). In this survey, data from a sample of incident cancer cases were collected from seven metropolitan statistical areas and two states in the U.S. The 1970 census provided denominators for incidence rates [28]. Histopathologic type-specific incidence rates were also calculated for all geographic areas combined in a later survey (1973–76) of five metropolitan statistical areas, five states, and the Commonwealth of Puerto Rico carried out as part of the Cancer Surveillance, Epidemiology, and End Results (SEER) Program by the Biometry Branch of the National Cancer Institute in conjunction with local medical institutions [61]. Several of the cities and one of the states were included in both the Third National Cancer Survey and the SEER Program Survey. Prior to these two surveys, the

Ten Cities Surveys of 1937 and 1947 analyzed data on a sample of leukemia cases (as well as on a sample of many other types of cancer cases) from ten U.S. cities: the leukemia cases were not further characterized by histopathologic type [62, 63].

The Third Cancer Survey attempted to be more representative than the two earlier surveys by selecting a population that included a substantial proportion of blacks (9.9% of all U.S. blacks), whites (10.4%), and other races (10.8%); its geographic distribution better represented the rural and farm areas than did the previous surveys.

An interesting study was conducted on all cases of leukemia diagnosed during 1956–67 in Metropolitan Atlanta (one of the cities included in the later Third National Cancer Survey) identified through hospital records, death certificates, and physician reports. Leukemia rates were roughly comparable for whites and blacks below age 70; beyond age 70 rates in whites continued to rise whereas those in blacks leveled off and declined. Each type of leukemia (acute, both types combined, CLL, and CML) was reduced in blacks to the same degree (60%–80% reduction). In younger blacks, acute leukemia rates were slightly lower (31%–33%) than among whites whereas no difference between whites and blacks was found for CML. The authors speculate that the greater apparent deficit of leukemia among elderly blacks was more likely to reflect socioeconomic factors, including reduced access to medical care rather than large racial differences in etiological factors per se [64].

ALL The overall mean annual age-, sex-, and race-adjusted incidence of acute lymphocytic leukemia from the Third Survey was 1.1 per 100,000, which is the same as the rate estimated from the SEER program. Age-adjusted incidence rates from the Third Survey and the SEER Program examined by sex and race indicated that males of all ages have higher rates than females. Age-specific rates among whites appear to be somewhat greater than those among nonwhites with exceptions for the age groups 10–14, 25–34, and 70–79 in which nonwhites have higher rates than whites. For all four race–sex groups, the age-specific incidence rates can be considered to have three main components: rapidly rising rates in early childhood with peaks at ages 1–4, then an antimode at ages 25–39, followed by a subsequent increase to a high of 4.5 per 100,000 in the 85+ age group (Figures 3-1, 3-2, and Table 3-5). (Age-specific incidence rates for blacks are based on small numbers and are therefore presented in tabular form rather than in graphical representation.)

Table 3-6 shows that ALL has a higher incidence rate among white males in Iowa when compared to other cities and states. (Rates for Iowan nonwhites are difficult to evaluate because they are based on a total of eight leukemia cases of all cell types, for all ages and both sexes.) Data on rural population subgroups in SEER cannot be compared because histopatho-

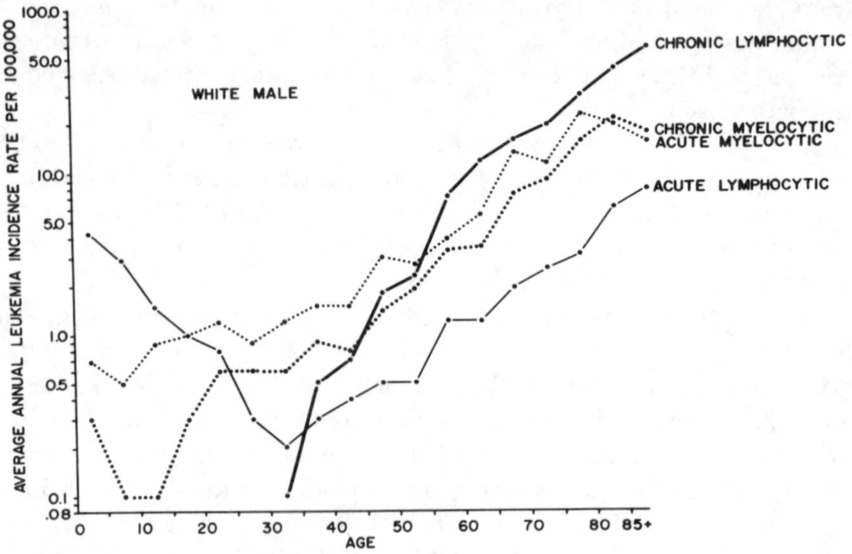

Fig. 3-1. Average annual age-specific leukemia incidence rates by cell type per 100,000 white males, United States, 1969–71. (Source: Third National Cancer Survey (1969–71) [55])

logic-specific rates were not separately determined for each metropolitan area and state included in the SEER Survey. However, in a number of studies, ALL has been found to be associated with rural areas and farming [65–74]. The lowest ALL incidence rates for both male and female whites occur in Detroit. Other cities included in the Third Survey did not exhibit similarly low rates (see Table 3-6).

CLL The annual age-, sex-, and race-adjusted U.S. incidence of CLL was estimated as 2.4 per 100,000 from the Third Survey and 2.6 per 100,000 from the SEER program. CLL incidence is 1.6 times higher in whites than nonwhites, a finding observed for all age groups and for both sexes at all ages except for nonwhites over 85, for whom incidence is greater in females. Possible explanations for the male predominance included the following: (1) males are more selectively exposed to certain occupational or other environmental risk factors: and (2) differential susceptibility of males and females to environmental factors due to hormonal, genetic, or other physiologic difference.

Between the ages of 25–29 and of 80–84, the CLL incidence rate increases 350-fold (see Figures 3-1, 3-2, and Table 3-5). The high CLL incidence in old age may be a result of a long-term exposure to an etiologic factor(s) necessary to the development of the disease. CLL may also be the result of an aging factor, which may reflect alterations in the immune system that are affected by various environmental carcinogens. The highest

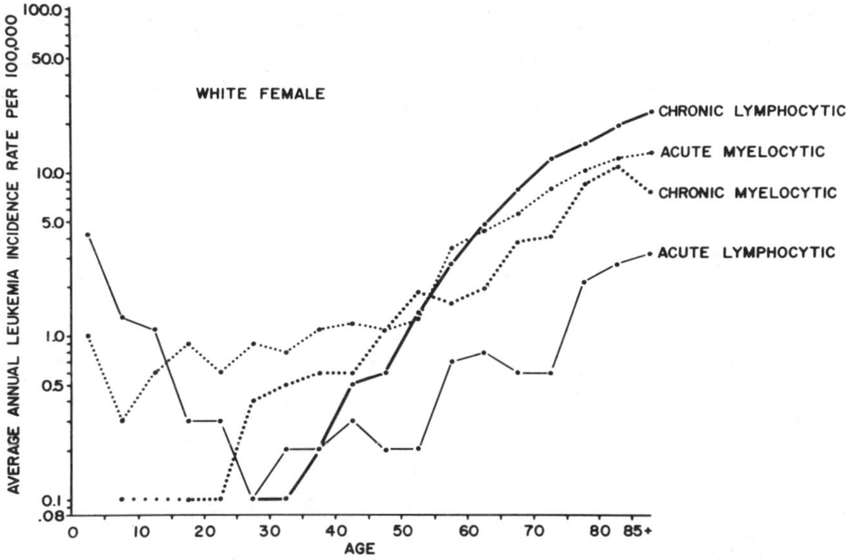

Fig. 3-2. Average annual age-specific leukemia incidence rates by cell type per 100,000 white females, United States, 1969–71. (Source: Third National Cancer Survey (1969–71) [55])

incidence rates for CLL in both sexes (4.1) were in Iowa; the high rates in whites, both males (5.1) and females (3.2), were primarily responsible. Among nonwhites, the highest rates (4.2) were noted in Minneapolis–St. Paul, particularly for males (6.3) (see Table 3-7).

ANLL ANLL consists of several types of acute leukemias as described earlier. All the acute nonlymphocytic leukemias (including AML, AMoL, AMMoL, etc.,) show trends by sex and race similar to those of ALL (Table 3-8). Whites have higher rates than nonwhites in those under 10 and those over 55; incidence rates in whites and nonwhites are approximately equal for the age range 10–54. Incidence is greater in males than in females for all age groups except for children under 5. Age-specific rates for both sexes and races combined show similar trends in each of the subgroups of ANLL. Age-specific rates by sex and race for all the nonlymphocytic leukemias combined are presented in Figures 3-1, 3-2, and Table 3-5. The age-specific incidence rates over a lifetime show a bimodal pattern with a small peak in children under 5 years of age and with a continuous increase in rates beginning at age 50. This pattern is similar for both sexes and races suggesting that risk factors for ANLL in children may be different than those in older individuals. The relatively low incidence of this type of leukemia in young adults may be due to a protective effect of hormonal and/or immunologic factors. The highest incidence of ANLL in males in both races was found in heavily industrial metropolitan areas

Table 3-5 Number of cases and average annual age-specific leukemia incidence rates by cell type per 100,000, black males and females, United States, 1969–71

Leukemia type	All ages[a]	<5	5–9	10–14	15–19	20–24	25–29	30–34	35–39
				Age group (years)					
				Males					
Acute lymphatic									
Number of cases	32	6	6	6	2	—	2	1	—
Rate	0.9	1.7	1.5	1.5	0.6	—	0.9	0.6	—
Chronic lymphatic									
Number of cases	70	—	—	—	—	—	—	—	1
Rate	2.9	—	—	—	—	—	—	—	0.6
Acute myelocytic									
Number of cases	54	1	1	3	5	1	2	3	2
Rate	2.0	0.3	0.3	0.7	1.5	0.4	0.9	1.7	1.2
Chronic myelocytic									
Number of cases	44	—	—	1	2	3	3	3	2
Rate	1.7	—	—	0.2	0.6	1.2	1.4	1.7	1.2
				Females					
Acute lymphatic									
Number of cases	18	3	4	5	1	1	—	—	—
Rate	0.5	0.8	1.0	1.2	0.3	0.3	—	—	—
Chronic lymphatic									
Number of cases	38	—	—	—	—	—	—	—	—
Rate	1.2	—	—	—	—	—	—	—	—
Acute myelocytic									
Number of cases	44	2	—	2	2	6	—	3	1
Rate	1.3	0.6	—	0.5	0.6	1.9	—	1.5	0.5
Chronic myelocytic									
Number of cases	35	1	—	2	1	—	—	1	5
Rate	1.2	0.3	—	0.5	0.3	—	—	0.5	2.5

[a]Summary type-specific rate for all ages is age-adjusted (1950 standard).
Source: Third National Cancer Survey [55].

including Pittsburgh and Detroit. Rates for females in both races were lower and fairly uniform in most areas, including the metropolitan (both industrial and nonindustrial cities) as well as rural areas. These sex-specific differences may suggest an etiologic influence of exposure to certain factors, perhaps occupational in nature; an occupational or other environmental factor specific to industrial metropolitan areas may also interact with male hormonal factors.

CML The annual incidence rate of CML adjusted for age, race and sex was estimated as 1.3 per 100,000 in the Third Survey and 1.4 per 100,000 by the SEER program. Like most of the other leukemias, this incidence is higher in males than females for whites of all age groups and for most age groups of nonwhites except for those ages 65–74. For each race–sex group except white males, the incidence rate increases gradually with age until

Table 3-5 (continued)

Age group (years)									
40–44	45–49	50–54	55–59	60–64	65–69	70–74	75–79	80–84	85+
				Males					
—	2	1	1	1	—	3	1	—	—
—	1.2	0.7	0.8	1.0	—	5.6	3.4	—	—
2	5	9	5	3	9	20	8	6	2
1.2	3.0	6.2	3.9	3.0	11.0	37.3	27.1	40.0	17.3
4	2	3	2	6	10	2	4	1	2
2.3	1.2	2.1	1.6	5.9	12.2	3.7	13.6	6.7	17.3
3	5	5	3	3	4	1	4	2	—
1.8	3.0	3.4	2.3	3.0	4.9	1.9	13.6	13.3	—
				Females					
—	1	—	—	—	1	1	1	—	—
—	0.5	—	—	—	1.0	1.5	2.5	—	—
—	4	6	3	1	8	7	1	4	4
—	2.1	3.6	2.1	0.9	8.1	10.7	2.5	18.5	21.8
2	5	4	4	3	5	1	—	2	2
1.0	2.6	2.4	2.8	2.6	5.1	1.5	—	9.2	10.9
2	4	3	—	2	7	4	1	2	—
1.0	2.1	1.8	—	1.7	7.1	6.1	2.5	9.2	—

age 35 and then increases more rapidly as shown in Table 3-9, though not as rapidly as CLL. White males have the same pattern except for a small peak in the under-5 age group due to "juvenile CML," a type of the disease with somewhat different characteristics (see Figures 3-1, 3-2, and Table 3-5).

No geographic differences are noted for CML. Rates are fairly uniform for both sexes and races (Table 3-9).

Age-adjusted leukemia incidence rates by histologic type, sex, and race are shown in Tables 3-10 and 3-11 from the Third Cancer Survey and SEER program, respectively.

SECULAR TRENDS

International

Until the late 1950's and early 1960's, it was not possible to examine time trends for the leukemias by histopathologic type. The first type-specific

Table 3-6 Acute lymphocytic leukemia (average annual incidence rates per 100,000 population by race, sex, and U.S. area, 1969–71)

	All races			White			Nonwhite		
Area	Total	Male	Female	Total	Male	Female	Total	Male	Female
Pittsburgh	1.3	1.6	0.9	1.4	1.7	1.0	0.4	0.8	—
Atlanta	1.1	1.4	0.8	1.2	1.4	1.0	0.9	1.6	0.2
Birmingham	0.9	1.3	0.6	1.0	1.6	0.4	0.9	0.7	1.1
Detroit	0.6	0.9	0.4	0.6	0.9	0.4	0.6	1.0	0.3
Iowa	1.8	2.4	1.1	1.8	2.5	1.1	1.0	—	2.0
Minneapolis–St. Paul	1.0	1.3	0.7	1.0	1.2	0.7	1.0	2.1	—
Dallas–Ft. Worth	1.3	1.3	1.2	1.3	1.4	1.3	1.0	1.3	0.8
Colorado	1.1	1.1	1.1	1.2	1.2	1.1	—	—	—
San Francisco–Oakland	1.2	1.3	1.0	1.2	1.4	1.0	0.8	0.8	0.8

Source: Third National Cancer Survey [55].

data were generated by the Registrar General of England and Wales, who obtained data on cell type for all persons dying of leukemia during 1945–57.

ALL During that time interval, an apparent increase in mortality due to ALL occurred in England and Wales. Further analysis suggested that this may have resulted from a change in classification, a reduction in the proportion of deaths coded as other and unspecified leukemias and a corresponding increase in those coded as ALL and AML [37].

Adelstein and White carried out a cohort analysis of leukemia (not char-

Table 3-7 Chronic lymphocytic leukemia (average annual incidence rates per 100,000 population by race, sex, and U.S. area, 1969–71)

	All races			White			Nonwhite		
Area	Total	Male	Female	Total	Male	Female	Total	Male	Female
Pittsburgh	2.0	2.7	1.3	1.9	2.6	1.4	2.4	3.8	1.2
Atlanta	1.7	2.1	1.2	1.8	2.4	1.2	1.4	1.4	1.4
Birmingham	1.9	2.4	1.5	1.8	2.4	1.2	2.1	2.3	2.0
Detroit	2.0	2.8	1.4	2.1	2.8	1.5	1.8	2.7	1.0
Iowa	4.1	5.1	3.3	4.1	5.1	3.2	3.1	4.2	2.0
Minneapolis–St. Paul	2.7	3.7	1.8	2.7	3.6	1.8	4.2	6.3	2.1
Dallas–Ft. Worth	2.2	2.5	1.9	2.3	2.5	2.1	1.4	2.3	0.6
Colorado	1.9	2.1	1.8	1.9	2.1	1.8	0.7	1.4	—
San Francisco–Oakland	2.0	2.6	1.5	2.3	2.8	1.7	1.3	1.7	1.0

Source: Third National Cancer Survey [55].

Table 3-8 Acute nonlymphocytic leukemia (average annual incidence rates per 100,000 population by race, sex, and U.S. area, 1969–71)

	All races			White			Nonwhite		
Area	Total	Male	Female	Total	Male	Female	Total	Male	Female
Pittsburgh	3.2	3.8	2.7	3.2	3.8	2.7	2.8	3.5	2.1
Atlanta	2.9	3.3	2.4	3.0	3.6	2.5	1.9	2.3	1.6
Birmingham	1.9	1.9	1.8	2.0	2.2	1.9	1.4	1.3	1.5
Detroit	3.3	4.1	2.5	3.5	4.3	2.6	2.6	2.9	2.3
Iowa	2.9	2.9	2.7	2.9	3.0	2.7	1.0	—	2.0
Minneapolis–St. Paul	2.6	2.9	2.2	2.7	2.9	2.3	2.1	4.2	—
Dallas–Ft. Worth	3.0	3.4	2.6	3.2	3.4	2.9	1.8	3.0	0.8
Colorado	2.7	2.9	2.5	2.6	3.0	2.4	4.2	1.8	6.1
San Francisco–Oakland	3.2	3.7	2.8	3.6	4.2	3.1	1.5	1.4	1.6

Source: Third National Cancer Survey [55].

acterized by histopathologic type) in children ages 0–4, who were born in England and Wales, Australia, and the United States between 1933 and 1963. (ALL comprises 60%–80% of leukemia in this age group in these countries.) Mortality steadily increased until approximately 1952 when it began to decline in all three countries. Rates during the entire period were lower in England and Wales than in Australia or the U.S. Analysis of mortality from other cancers between 1933 and 1963 in children ages 0–4 showed a similar increase until 1952 and then a decline. Although a detailed survival analysis was not performed, the investigators found fewer

Table 3-9 Chronic myelocytic leukemia (average annual incidence rates per 100,000 population by race, sex, and U.S. area, 1969–71)

	All races			White			Nonwhite		
Area	Total	Male	Female	Total	Male	Female	Total	Male	Female
Pittsburgh	1.2	1.6	0.9	1.1	1.6	0.8	2.0	2.0	2.0
Atlanta	0.9	1.2	0.6	1.0	1.2	0.7	0.9	1.4	0.4
Birmingham	1.2	1.5	0.9	1.3	1.3	1.2	1.1	2.0	0.3
Detroit	1.4	2.0	1.0	1.4	2.1	0.9	1.3	1.4	1.2
Iowa	1.8	2.1	1.5	1.8	2.1	1.5	1.0	—	2.0
Minneapolis–St. Paul	1.6	1.6	1.7	1.6	1.6	1.6	2.1	2.1	2.1
Dallas–Ft. Worth	1.4	1.5	1.3	1.4	1.5	1.4	1.0	1.5	0.6
Colorado	1.4	1.7	1.1	1.3	1.7	1.0	3.0	1.9	4.2[a]
San Francisco–Oakland	1.1	1.4	0.8	1.1	1.4	0.8	1.2	1.4	1.0

[a]Based on only 3 cases.
Source: Third National Cancer Survey [55].

Table 3-10 Age-adjusted leukemia incidence rates per 100,000 per year by histologic type, sex, and race, 1969–71

	Males			Females		
	Whites	Nonwhites	Total	Whites	Nonwhites	Total
ALL	1.5	1.0	1.5	0.9	0.5	0.9
ANLL	3.6	2.9	3.5	2.5	1.7	2.4
Acute granulocytic	2.7	2.0	2.6	1.8	1.3	1.7
Monocytic	0.4	0.3	0.4	0.3	0.1	0.3
Other acute leukemia	0.5	0.6	0.5	0.4	0.3	0.4
CLL	3.0	2.9	3.0	1.5	1.2	1.9
CML	1.9	1.8	1.9	1.1	1.2	1.1

Source: Third National Cancer Survey [55].

cases in children up to age 14 after 1952, implying that there was a reduction in incidence and not simply a postponement of death. Survival analysis would be helpful in order to understand the possible contribution of therapy [75].

Court-Brown and Doll investigated trends of childhood and adolescent leukemia death rates, 1945–59, by cell type. They found increasing mortality rates in children with ALL and AML for each age group and ascribed some of this to improved diagnosis of leukemia and better treatment for infections that formerly might have caused death before leukemia was diagnosed [6]. However, a substantial part of the increase was believed due to changes in classification in which a larger proportion of deaths were diagnosed as ALL or AML during the latest time period while a smaller proportion were diagnosed as unspecified or unclassified.

CLL Between 1945–49 and 1955–57, mortality from CLL increased in males above age 45 but not in younger age groups in England and Wales.

Table 3-11 Age-adjusted leukemia incidence rates per 100,000 per year by histologic type, sex, and race, 1973–76

	Males			Females		
	Whites	Blacks	Total	Whites	Blacks	Total
ALL	1.4	0.5[a]	1.4	0.9	0.7[a]	0.9[b]
ANLL	3.9	2.7	3.9	2.6	2.4	2.6
CLL	3.8	3.8[a]	3.8	1.9	1.4[a]	1.8
CML	1.8	1.8[a]	1.8	1.0[b]	1.1[a]	1.1

[a]Standard error of the rate is 10% or greater.
[b]Standard error of the rate is between 5% and 10%.
Source: SEER Program Survey [60].

The increase in mortality in women between the two time periods was somewhat less than that for men, but was most apparent in those 60 years and older. This differential increase in rates between men and women is reflected in a change in the male to female ratio from 1.7:1 in 1945–49 to 2.0:1 in 1955–57. Since mortality in younger persons with CLL did not increase between the three time periods, Court-Brown and Doll ascribed the increases at older ages to improvements in diagnosis, increased access to medical care, and changes in treatment for other conditions that resulted in improved survival in older persons [28].

ANLL In an analysis of data collected between 1945 and 1958 from the Registrar General's Office for England and Wales, the General Registry Office for Scotland, the U.S. National Office of Vital Statistics, and the Canadian Dominion Bureau of Statistics, Lee noted that acute leukemia rates, both ALL and ANLL combined, fell beginning at ages 3–4 in children and were at lowest levels between ages 10 and 35 before beginning to rapidly increase. Further analysis of leukemia subtypes in this age group showed a small peak between ages 15 and 19, for males only, that was found to be due to AML. The pattern was similar for the four countries included in the analysis (England and Wales, Scotland, Canada, and the United States). Time trend analysis indicated that in England and Wales, this small AML age- and sex-specific peak in adolescent males was present as far back as 1911–15 [37].

Geary et al. noted a twofold increase in incidence for acute myelogenous leukemia (from 1.24 to 2.60 per 100,000 per year) in Lancashire, England, between 1965–70 and 1971–76 [76]. The increase was not specific to one particular town or hospital cachement area, but was present throughout the entire area of Lancashire. However, further analysis by age and sex was not reported by these investigators. The rise in AML incidence was also noted throughout England and Wales between the two time periods.

A similar increasing incidence of AML was noted in Sweden by Brandt et al. between 1964 and 1973 [77]. Age- and sex-specific analysis showed that the increase was most apparent in men 50 years and older with a slight increase in men, 20–34 years of age. The increase among females was much less pronounced, suggesting that the increase was not due to improvements in diagnosis or more complete registration. Kemp et al. reported 1½-fold increase in AML in Scotland between 1968–72 and 1973–77 [78]. However, the increase occurred at all ages, both sexes, in all social classes and in all parts of the country. During that period, the total number of cancers from all sites registered in Scotland rose by only 10%, a much smaller increase than the 150% for AML.

CML Gould et al. analyzed 169 case of CML admitted to hospitals in Edinburgh and Aberdeen, Scotland, during 1938–51, with an average age of 49.8 and a sex ratio (males to females) of 0.9:1. The number of cases of CML admitted to these hospitals increased 60% over the time period of the study [79].

For England and Wales, Court-Brown and Doll noted an increased mortality from CML in males ages 50 and older between 1945–49 and 1955–57. A similar but smaller increase was noted for women 60 and older. The ratio for male to female mortality was 0.85 in 1945–49 and 1.2 in 1955–57 [28]. Geary et al. also noted a twofold increase in CML in Lancashire, England, between 1965–70 and 1971–76 [76]. The reason for this was not apparent since rates for CML appeared constant between the two time periods for the rest of England and Wales. CML did not increase in incidence in Scotland as much as did AML between 1968–72 and 1973–77 [78]. The rates for CML increased approximately 10%, similar to trends for all other cancers.

Developing Countries

A number of studies have reported some interesting secular trends for the leukemias in developing countries, but the data either were not population-based, or were not histopathologic type-specific. Nevertheless, since these studies often represent the only recent information about leukemia trends in these countries, the findings are briefly summarized.

Kim and Lee noted that the number of cases of acute leukemia (including ALL and AML) seen at Seoul National University Hospital, Korea, increased more than threefold between 1955–59 and 1969–71 while the relative frequency of chronic leukemias decreased over the same time period [80]. As in Japan, China, and India, CLL is very rare in Korea, comprising only 5% of the leukemia cases; AML is the most common (38.1%). Since the age-, sex-, and cell-specific secular trends were not analyzed, it was impossible to determine whether the apparent increase in leukemia was due to better ascertainment, referral, and/or registration of cases, or represented a true increase. Because a population-based approach was not used, it is difficult to interpret the findings.

Kushwaha et al. noted that the incidence of leukemia in Lucknow, Uttar Pradesh, India, increased from 1970 through 1976, but these authors did not analyze the increase by cell type [81]. However, CML was the most frequent morphologic type of leukemia (56.1%), followed by AML (31.6%) and ALL (8.1%). CLL (4.2%) is quite uncommon throughout India. Most acute leukemia cases resided in urban communities whereas most of the chronic leukemia cases were from rural areas. In addition, 25%

of the CML cases were homemakers, another 20% were farmers. Approximately half of those with CLL were farmers.

Booth and Amato noted an increase in reported cases of leukemia in Papua, New Guinea, between 1958–67 and 1968–76; the country has had a Central Tumor Registry since 1958 [82]. However, these authors believed that the increase, mainly in the 25–44 year age group, could be almost entirely explained by improvements in medical services. Between the two time periods, childhood ALL frequency rose from 17% to 27% and childhood AML declined from 50% to 40%. The remaining proportions of childhood leukemias are designated "unclassified." The greater frequency of AML compared with ALL is similar to the frequency noted in most African countries, and, as in Africa, leukemias are less common than lymphomas in the 0–14 age group. The low frequency of CLL (2%) resembles that found in Japan, China, India, and Korea. But the age distribution of the population must also be considered since the average life expectancy for a native of Papua, New Guinea, is 50 years. Over 25% of those with CML were under age 20 at the time of diagnosis, similar to the African findings.

United States

All leukemia types combined Prior to the 1960's, most analyses of secular trends of U.S. leukemia incidence and mortality did not consider histopathologic types. Data will first be presented from several early reports of trends for leukemias as a group. Other early studies of either hospital-based or population-based data for groups of histologic types will also be discussed. Finally more recent, histopathologic type-specific analyses of secular trends will be described.

Gilliam and Walter examined leukemia mortality trends (for all histopathologic types combined) in the United States for 1921–55 and noted a continuous increase in leukemia mortality rates until about 1940 [7]. MacMahon noted that the increase in mortality during 1920–40 was greatest in those areas in the U.S. which initially had the lowest rates from leukemia, including Vermont, New Hampshire, Kentucky, Arkansas, Oklahoma, and Idaho [48]. Similar patterns were noted in data from the New York [83] and Connecticut [84] cancer registries. The reasons for the increase were not completely understood, but they probably included improvements in diagnostic tools, increased access to medical care, and changes in death certification and registration. Since histopathologic type-specific data are not available for that time period, it is impossible to know whether the increase involved all or only a limited number of cell types.

In 1940, the rate of increase began to diminish in all age groups except for persons over 75 years of age, but most notably in the youngest age groups [7]. This decline was particularly striking in infants under 1 year of age. In 1945, mortality rates also began to decrease in 1-year-olds, and in 1950, in 2-year-olds [8, 85].

Fraumeni and Miller reported that the decline in increase in mortality rates for leukemia leveled off beginning about 1955 in both white and nonwhite populations in both the United States and England and Wales [85, 86]. During 1961–65, there was actually a downturn in rates among the U.S. white population but not among the nonwhites nor the population in England and Wales. Among U.S. whites, it affected persons ages 1–74, but was most pronounced in the group 1–4 years of age. Analysis by cell type (ICD 7th Revision) showed a decrease in rates for both lymphatic and myeloid leukemias in all age groups under 75 years of age. Acute leukemia (a separate designation in the 7th ICD Revision) declined in age groups under 45 years of age but increased in older groups.

In an analysis of leukemia incidence in Olmstead County, Minnesota, for 1935–64, Kyle et al. noted no significant overall increase during this time period. However, they noted a doubling in the rates of those 75 years and older between 1935–44 and 1955–64 and a slight decrease in the rates for those under 20 years of age. They believed that the stability of rates in Olmstead County might be a better indicator of leukemia trends than were national or tumor registry data because the method of ascertainment of cases in Olmstead County was more complete; it did not change over the decades studied; and all pathologic material, including peripheral smears, bone marrow aspirates, and biopsies, from all cases were reviewed by one hematologist. The increase in rates in the elderly may be due to greater longevity in the elderly—a result of increased care and improvements in the treatment of stroke, heart disease, and other competing causes of death [87].

Cell-type-specific trends There are very few population-based studies that have separately examined secular trends of the specific leukemia subtypes, particularly incidence rates. Most studies of time trends have analyzed leukemia mortality, especially survivorship using cases from a single hospital or group of hospitals in which the cases may represent a highly selected group of leukemia patients.

Although mortality rates may represent a fairly accurate proxy measure of incidence for a number of types of cancer, the analysis of mortality data particularly over the past two decades may be quite misleading for ALL and more recently, ANLL, since survivorship has dramatically improved for both of these leukemia subtypes.

The few population-based incidence studies have suggested some promising leads for epidemiologic studies of risk factors. However, the existing studies have focused mostly on ALL and ANLL and sometimes only upon a limited age group within the particular leukemia subtypes. Trends in incidence rates of CLL and CML have received virtually no attention. Changing trends in age at diagnosis, sex ratio changes, as well as changes in socioeconomic characteristics of cases, urban–rural differences, and relationship to occupations, industries, and other geographically defined environmental exposures need further attention.

ALL Although in 1920 an early childhood peak in ALL began to appear in U.S. children ages 3–4, Fraumeni and Miller noted that after 1940 a more pronounced increase in leukemia mortality occurred in U.S. white children in this age group. This age peak, also noted after 1910 in England, became more pronounced with time and may have been primarily due to changing patterns of survival in young children with ALL, namely a decline in mortality in children with ALL under age 2. As survival increased in children with onset of ALL at ages 1–2, death was postponed to ages 3–4. Initially no such peak was found among U.S. nonwhites or Japanese children. However, in 1960–64, a suggestion of a peak began to occur among 4-year-old U.S. nonwhites and 3-year-old Japanese [6, 86].

However, as indicated in Table 3-12, population-based secular trends studies carried out in Olmstead County, Minnesota (1935–74) [88] and Baltimore SMSA (1960–74) [89, 90] showed no significant change in ALL incidence. But hospital-based studies by Cutler et al. (1940–62) [91] and Zippin et al. (1955–69) [92] did demonstrate an improvement in median survival over the years studied (Table 3-12). The Zippin et al. study suggests that this difference may, in part, be explained by two findings:

1. During the earlier period, a higher proportion had more severe disease at time of diagnosis whereas during the later period, cases may have been detected at an earlier stage in the natural history of ALL.
2. During the later period, a higher proportion of cases were treated with combination chemotherapy.

A population-based study carried out by Szklo et al. demonstrated a marked increase in 2-year survival between 1960–64 and 1965–69 for both whites and blacks as indicated in Table 3-12, but further improvement in 1970–74 could only be documented for whites [90].

Other interesting trends are noted in Table 3-12. The proportion of

Table 3-12 Secular trends in ALL incidence, mortality, and survival, United States, 1913–74

Author and date	Source and number of cases	Time periods of study	Leukemia incidence or mortality	Leukemia survival	Leukemia case characteristics	Treatment differences	Other differences
Shimkin et al. 1954 [94]	University of California Hospital 149 cases	1913–29 1930–39 1940–47		No significant differences in median survival (4.5 ± 0.36 months) Longer survival in all 3 periods associated with initially lower leukocyte counts	No significant differences by period for: Age of onset (42% diagnosed before age 5; 66% before age 10) Male to female ratio 1.6:1	Survival was not affected by treatment used	
Cutler et al. 1967 [91]	98 U.S. hospitals participating in the End Results Evaluation Program 256 cases	1940–49 1950–54 1955–59 1960–62		Median survival: 1940–49 was 2–3 months 1960–62 was 11 months with females experiencing a longer survival Longer survival associated with initially lower leukocyte counts	No significant differences by period for: Male to female ratio 1.2:1 Proportion classified as ALL: 70% in all four time periods		Proportion of cases with bone marrow diagnosis: 1940–49 47% 1950–54 62%–72% 1955–59 62%–84% 1960–62 79%
Fraumeni et al. 1971 [15]	Children's Cancer Research Foundation 527 cases	1947–53 1954–59 1960–65		Median survival over the entire time period was 13 months (time trends not examined)	Male to female ratio 1.08:1 Peak in frequency occurred at age 3½		Proportion of cases diagnosed as lymphocytic: 1947–53 37.9% 1960–65 50.1% Proportion of cases diagnosed as undifferentiated: 1947–53 32.0% 1960–65 15.4%

Zippin et al. 1973 [92]	647 cases diagnosed 1955–64 at 20 hospitals participating in the End Results Evaluation Program 460 cases diagnosed 1965–69 at 33 hospitals in the End Results Evaluation Program	1955–64 1964–69		Median survival: 1955–64 9.5 months 1965–69 16.8 months	Male to female ratio: 1955–64 1.5:1 1965–69 1.1:1 No difference in age distribution by period	Higher proportion were classified as severely disabled and had less favorable prognoses in the earlier time period Higher proportion had combination chemotherapy during the later time period
Linos et al. 1978 [88]	Olmstead County, Minnesota; population-based cases 42 cases	1935–44 1945–54 1955–64 1965–74	No significant change in incidence rates over the 4 time periods in males or females			
Szklo et al. 1978 [90] Gordis et al. 1980 [89]	Baltimore SMSA 278 cases	1960–64 1965–69 1970–74	Age-adjusted incidence rates per 100,000 (ages 0–19): 1960–64 Whites 2.0 Blacks 1.0 1965–69 Whites 2.3 Blacks 0.7 1970–74 Whites 2.6 Blacks 0.9	Age-adjusted 2-year survival rates (ages 0–19): 1960–64 Whites 32% Blacks 25% 1965–69 Whites 47% Blacks 59% 1970–74 Whites 79% Blacks 59%	77% of all leukemia cases in Baltimore SMSA 1960–74 had ALL Peak incidence in white children with ALL was in the age group 1–4 but a similar peak was not found in blacks	

cases in which bone marrow specimens were employed in diagnosis increased substantially between 1940–49 and 1960–62 as found by Cutler et al. [93]. Fraumeni et al. [15] showed that the proportion of childhood leukemia cases diagnosed as lymphocytic increased from 37.9% in 1947–53 to 50.1% in 1960–65 whereas the proportion diagnosed as undifferentiated decreased from 32.0% in 1947–53 to 15.4% in 1960–65.

No consistent differences were noted during the time periods indicated in Table 3-12 in the male to female sex ratios which ranged from 1.08–1.6:1. The studies summarized in Table 3-12 showed no definite trends in age of onset of ALL over the different periods studied. An interesting clinical finding noted by Shimkin et al. [94] and Cutler et al. [93] was that longer survival was associated with initially lower leukocyte counts; this finding was noted over all time periods studied by these investigators.

CLL As noted previously, Court-Brown and Doll in England and Wales [28, 29] and Keogh and McCall [30] in Australia found an increase in CLL mortality in later time periods (the 1950's) in older males and females compared with earlier time periods (the 1940's). This was believed to parallel an increase in incidence in age groups over age 65 due in part to improvements in diagnosis as well as to improvements in survival as a result of reduction in competing causes of death. As depicted in Table 3-13 in an analysis of secular trends in survival in 98 U.S. hospitals participating in the End Results Evaluation Program, Cutler et al. showed substantial improvements in median survival from CLL between 1940–49 and 1955–59 with a slight downturn in the latest period studied (1960–62) for all age groups and both sexes. They also noted that survival, in general, was better in females than males [91]. Other investigators found little variation in median survival between 1913–29 and 1940–47 which fell within a fairly narrow range of 33–43 months (Table 3-13). A population-based study in Olmstead County, Minnesota, showed no change in CLL incidence between 1935–44 and 1965–74 [87, 88]. However, there was a change in sex ratio: The male to female ratio increased from 2.1:1 in the earliest time period to 3.5:1 in the latest period studied (Table 3-13).

Shimkin et al. found an increase in median age of onset from 52 years in 1913–29 to 58 years in 1940–47 as noted in Table 3-13. This finding was consistent with the increased incidence in older age groups over time as described above. Shimkin et al. as well as others were unable to evaluate the role of treatment in survival because of great variation in types of treatment employed, duration and dose of treatment used, and types and sites of radiation given as treatment [94].

ANLL Of all the leukemia types, changing time trends for ANLL appear to suggest the most promising clues for epidemiologic, particularly etiologic, studies. However, it must be noted that, for the two population-based studies, the numbers are small and hence results must be interpreted with caution. Linos et al. [88] noted a marked 4½-fold increase in incidence in males ages 50 and older between 1940 and 1970 (Table 3-14). Since all cases had blood smears and bone marrows reviewed by the same hematologist, the cases are homogeneous (within the spectrum of leukemias designated as ANLL) but it is possible that the apparent increase could be due to improvements in diagnosis. This is somewhat unlikely because a similar increase was not found for females and hence the substantial increase in males needs to be explained.

The findings of Gordis et al. [89] of the significantly increased incidence in black children, particularly those in the high SES group, between 1960 and 1974 are also intriguing. They also carefully examined the possibility that the finding might be artefactual and finally concluded that the change was probably real and might be related to environmental factors. This conclusion also deserves additional investigation.

Studies by Cutler et al. [91] and Fraumeni et al. [15] as summarized in Table 3-14 suggest improvements in survival between 1940 and 1965, most likely due to the development of improved supportive as well as new chemotherapy treatments. Of course, the most dramatic developments in extending survival and dramatically increasing the proportion of cases achieving remission have occurred in the last few years.

Changes in the male to female ratios have not been consistently found in all studies. Other problems in interpreting trends are due to major changes in both histopathologic classification and ICD classification that have particularly affected leukemias grouped under the generic title of ANLL.

CML Very little information is available about secular trends involving this leukemia subtype. A population-based study examining U.S. incidence patterns for CML from 1935 to 1974 carried out in Olmstead County, Minnesota, showed a decrease in incidence between the time period 1935–1944 and subsequent time periods (Table 3-15). The total number of cases was very small and therefore rates may not be stable. Small numbers also precluded useful analysis of age- and sex-specific rates [88]. A large survival analysis carried out by Cutler et al. [91] and summarized in Table 3-15 showed inconsistent changes in survival experience over the time period of the study except for males in both the 45–64 and 65–74 year age groups whose median survival improved from 7 months to 16 months and

Table 3-13 Secular trends in CLL incidence, mortality, and survival, United States, 1913–74

Author and date	Source and number of cases	Time periods of study	Leukemia incidence or mortality	Leukemia survival	Leukemia case characteristics	Treatment differences	Other differences
Shimkin et al. 1954 [94]	University of California Hospital 137 cases	1913–29 1930–39 1940–47		Median survival was 36 ± 2.7 months Age at onset not associated with survival Initial leukocyte count not associated with survival	Mean age of onset: 1913–29 52 years 1940–47 58 years Male to female ratio 2.5:1	Not evaluable because of variation in: Type of radiation Duration of treatment Total dosages Sites of administration of treatment	
Cutler et al. 1967 [93]	98 hospitals participating in End Results Evaluation Program 3061 cases	1940–49 1950–54 1955–59 1960–62		Median survival for males 45–64: 1940–49 16 months 1950–54 32 months 1955–59 48 months 1960–62 44 months A similar reverse in survival occurred in 1960–62 involved all age groups and both sexes Survival in general was better in females than in males			
Minot and Isaacs 1924 [95]	C. P. Huntington Memorial Hospital 98 cases	Not stated		Median survival: 3.45 years—same in those treated with radiation and those not treated	Male to female ratio 2.9:1		

Leavell 1938 [96]	Massachusetts General, New York, University of Pennsylvania Hospitals 128 cases	1917–36		Median survival: 3.6 years	Male to female ratio 2.6:1 Peak age of onset was 45–55 years
Zippin et al. 1973 [97]	24 hospitals participating in End Results Evaluation Program 839 cases	1955–64 (cutoff date for follow-up was 1969)		Five year survival: 34% For each sex, survival decreased with increasing age Survival negatively associated with: Bleeding Infection Poor performance status Lymphadenopathy Hematologic abnormalities Pathologic bone marrow findings	
Osgood 1964 [98]		1941–54		Five year survival: 37%	
Boggs et al. 1966 [99]		1945–64		Five year survival: 54%	
Linos et al. 1978 [88]	Olmstead County, Minnesota, population-based cases 69 cases	1935–44 1945–54 1955–64 1965–75	No significant change in incidence over the four time periods		Male to female ratio: 1935–44 2.1:1 1965–74 3.5:1

Table 3-14 Secular trends in ANLL incidence, mortality, and survival, United States, 1935–74

Author and date	Source and number of cases	Time periods of study	Leukemia incidence or mortality	Leukemia survival	Leukemia case characteristics	Treatment differences	Other differences
Cutler et al. 1967 [93]	42 Connecticut hospitals participating in the End Results Evaluation Program 108 cases of AML	1940–62		Mean survival for children ages 0–9 at diagnosis 1940–49 1.5 mos. 1957–62 10.0 mos. Mean survival for children ages 10–19 at diagnosis did not change over time period. Average was 2.6 mos.	Male to female ratio 1.6:1		
Fraumeni et al. 1971 [15]	Children's Cancer Research Center, Boston 378 cases of ANLL including: AML AMoL "undifferentiated" leukemia	1947–53 1954–59 1960–65		Median survival AML 6.9 mos. Median survival AMoL 3.3 mos.	AML—Male to female ratio: 1947–53 2:1 1960–65 1.42:1 AMoL—Male to female ratio: 1947–53 1:1 1960–65 1.25:1		

Linos et al. 1978 [88]	Olmstead County, Minnesota population-based cases 75 (AML) cases	1935–44 1945–54 1955–64 1965–74	ANLL incidence in males ages 50 and over increased 1940– 5.0/100,000 per year 1960– 10.3/100,000 1970– 22.6/100,000		Male to female ratio increased over the time period: 1935–44 1.3:1 1965–74 2.4:1	For the entire time period blood smears and bone marrows were reviewed by the same hematologist
Gordis et al. 1980 [89]	Baltimore SMSA 58 cases	1960–64 1965–69 1970–74	Age-adjusted incidence rates per 100,000 (ages 0–19): 1960–64 Whites 0.5 Blacks 0.2 1965–69 Whites 0.6 Blacks 0.3 1970–74 Whites 0.7 Blacks 1.1	Age-adjusted 1-year survival rates (ages 0–19) for whites (numbers of blacks too small to analyze): 1960–64 33% 1965–69 42% 1970–74 71%	Peak age of onset in children was 2–6 in both whites and blacks Male to female ratio: Whites 1.3:1 Blacks 0.9:1	

Table 3-15 Secular trends in CML incidence, mortality, and survival, United States, 1940–62

Author and date	Source and number of cases	Time periods of study	Leukemia incidence or mortality	Leukemia survival				Leukemia case characteristics	Treatment differences	Other differences
				Years	Age	M Median survival (months)	F			
Cutler et al. 1967 [91]	98 U.S. hospitals participating in the End Results Evaluation Program 2060 cases	1940–49 1950–54 1955–59 1960–62		1940–49	20–44	14	24			
					45–64	7	19			
					65–74	7	2			
				1950–54	20–44	18	22			
					45–64	7	14			
					65–74	4	11			
				1955–59	20–44	16	28			
					45–64	12	21			
					65–74	9	10			
Linos et al. 1978 [88]	Olmstead County, Minnesota, population-based cases 12 cases	1935–44	1.3							
		1945–54	0.6							
		1955–64	0.6							
		1965–74								

7 to 14 months, respectively. Overall survival for females in both 20–44 and 45–64 year age groups was longer than that for males.

REFERENCES

1. J. Clemmesen, "On the epidemiology of leukemia," in *Advances in Acute Leukemia,* pp. 1–50 (F.J. Cleton, D. Crowther, and J.S. Malpas, eds). Amsterdam: North-Holland; New York: American Elsevier, 1974.
2. M. Alderson, "The epidemiology of leukemia," *Adv. Cancer Res.* **31:**1–76 (1980).
3. M.R. Alderson, "Central government routine health statistics," in *Reviews of United Kingdom Statistical Sources,* Vol. 2 (W.F. Maunder, ed). London: Heinemann, 1974.
4. M. Segi and M. Kurihara, *Cancer Mortality for Selected Sites in 24 Countries (1962–63),* No. 4, pp. 1–358, Sendai, Japan: Department of Public Health, Tohoku University, School of Medicine, 1966.
5. M. Segi, H. Noye, and R. Segi, *Age Adjusted Death Rates for Cancer for Selected Sites (A-Classification) in 43 Countries in 1972.* Nagoya, Japan: Segi Institute of Cancer Epidemiology, 1977.
6. W.M. Court-Brown and R. Doll, "Leukemia in childhood and young adult life. Trends in mortality in relation to aetiology," *Br. Med. J.* **1:**981–988(1961).
7. A.G. Gilliam and W.A. Walter, "Trends of mortality from leukemia in the United States, 1921–55," *Pub. Health Rep.* **73:**773–784 (1958).
8. J.C. Slocumb and B. MacMahon, "Changes in mortality rate from leukemia in the first five years of life," *N. Engl. J. Med.* **268:**922–925 (1963).
9. D. Mainwaring, "Epidemiology of acute leukaemia of childhood in the Liverpool area," *Br. J. Prev. Soc. Med.* **20:**189–194 (1966).
10. G. Hems and A. Stuart, "Childhood leukemia in Scotland, 1939–68," *Scott. Med. J.* **17:**13–17 (1972).
11. D.J. Klaassen, "Acute leukemia in children in the province of Saskatchewan, 1961–66," *Can. Med. Assoc. J.* **101:**87–90 (1969).
12. S.S. Meighan, "Leukemia in children. Incidence, clinical manifestations and survival in an unselected series," *Cancer* **16:**656–664 (1963).
13. H. Nishiyama, "Relative frequency and mortality rate of various types of leukemia in Japan," *Gann* **60:**71–81, 343–346 (1969).
14. R.W. Miller, "Cancer epidemics in the People's Republic of China," *J. Natl. Cancer Inst.* **60:**1195–1203 (1978).
15. J.F. Fraumeni, Jr., M.D. Manning, and W.J. Mitus, "Acute child-

hood leukemia: epidemiologic study by cell type of 1,263 cases at the Children's Cancer Research Foundation in Boston, 1947–65," *J. Natl. Cancer Inst.* **46**:461–470 (1971).

16. N.C. Allan and E.J. Watson-Williams, "A study of leukemias among Nigerians in Ibadan," in *Proc. 9th Cong. Eur. Soc. Haematol.*, pp. 906–915. Basel: S. Karger, 1963.
17. G.M. Edington and M. Hendrickse, "Incidence and frequency of lymphoreticular tumors in Ibadan and the Western State of Nigeria," *J. Natl. Cancer Inst.* **50**:1623–1631(1973).
18. E.M. Essein, "Leukaemia in Nigerians. I. The acute leukaemias," *Afr. J. Med. Sci.* **3**:117–130(1972).
19. A.F. Fleming, "Leukaemia in the guinea savanna of northern Nigeria," in *Advances in Comparative Leukemia Research*, pp. 53–54 (P. Bentvelzen, J. Hilgers, and D.S. Yohn, eds). Amsterdam: Elsevier/North Holland Biomedical Press, 1978.
20. A.F. Fleming. "Epidemiology of the leukaemias in Africa," *Leuk. Res.* **3**:51–59 (1979).
21. S. Amsel and J.S. Nabembezi, "Two-year survey of hematologic malignancies in Uganda," *J. Natl. Cancer Inst.* **52**:1397–1401 (1974).
22. E.G. Kasili and J.R. Taylor, "Leukemia in Kenya," *E. Afr. Med. J.* **47**:461–468(1970).
23. F. Lothe, "Leukemia in Uganda," *Trop. Geogr. Med.* **19**:163–171 (1967).
24. G.T. O'Conor and J.N. Davies, "Malignant tumors in African children: with special reference to malignant lymphoma," *J. Pediatr.* **56**:526–535 (1960).
25. J. Sonnet, J.L. Michaux, and C. Hekster, "Incidence and forms of leukemia among Congolese Bantus," *Trop. Geogr. Med.* **18**:272–286 (1966).
26. C. Jeffrey and M. Gelfand, "Leukaemia in the Rhodesian African," *J. Trop. Med Hyg.* **75**:176–179 (1972).
27. J.N. Davies, J. Knowelden, and B.A. Wilson, "Incidence rates of cancer in Kyadondo County, Uganda, 1954–60," *J. Natl. Cancer Inst.* **35**:789–821 (1965).
28. W.M. Court-Brown and R. Doll, "Adult leukemia," *Br. Med. J.* **1**:1063–1069 (1959).
29. W.M. Court-Brown, R. Doll, and I.D. Hill, "Leukemia in Britain and Scandinavia," *Pathol. Microbiol.* **27**:644–654 (1964).
30. E.V. Keogh and C. McCall, "Mortality from leukemia in Victoria, 1946 to 1955: a report from the Central Cancer Registry, Melbourne," *Med. J. Aust.* **2**:632–634 (1958).
31. W. Haenszel and M. Kurihara, "Studies of Japanese migrants. I.

Mortality from cancer and other diseases among Japanese in the United States," *J. Natl. Cancer Inst.* **40:**43–68 (1968).

32. F.W. Gunz, "Studies on the incidence and aetiology of leukaemia in New Zealand," *N.Z. Med. J. Haematol. Suppl.* **65:**857–864(1966).
33. A.C. Templeton, "Leukaemia," *Recent Results Cancer Res.* **41:**298–301 (1973).
34. I. Kela-We, "Les leucémies lymphoides chroniques au Zäire: á propos de 39 cas," *Med. Afr. noire* **24:**255–259 (1977).
35. E.M. Essein, "Leukemia in Nigerians. II. The chronic leukaemias," *E. Afr. Med. J.* **53:**96–103 (1976).
36. D.R.W. Haddock, "The pattern of leukemia in Accra, Ghana," *J. Trop. Med. Hyg.* **70:**60–62 (1967).
37. J.A.H. Lee, "Acute myeloid leukemia in adolescents," *Br. Med. J.* **1:**988–992 (1961).
38. H. Nishiyama, "Childhood leukemia mortality in Japan, 1947–1966: with special reference to the group of under five years of age," *Gann* **60:**569–581 (1969).
39. R.F. Lowe, "The incidence of leukemia in the Rhodesian African—a five-year hospital survey." *Cent. Afr. J. Med.* **20:**80–84 (1974).
40. B. Kusman, R.J. Jacobson, and L.G. MacDougall, "Childhood and adult acute leukemia in Johannesburg blacks," *South Afr. Med. J.* **54:**1007–1010 (1978).
41. H. Barsoum, "A case of chloroma," *Br. Med. J.* **1:**282–283 (1938).
42. T. Fahmi and M. Kinawi, "A case of chloroma," *J. Egypt Med. Assn.* **34:**202–205 (1951).
43. M. Tomonaga, "Statistical investigation of leukemia in Japan," *N.Z. Med. J. Haematol. Suppl.* **65:**863–869 (1966).
44. J. Waterhouse, C. Muir, P. Correa, and J. Powell (eds), *Cancer Incidence in Five Continents,* Vol. III, Sci. Publ. No. 15. Lyon, France: IARC, 1976.
45. M.N. Lowenthal, "Chronic myeloid leukemia in Zambians," *Trop. Geogr. Med.* **27:**132–136 (1976).
46. W.A. Walter and A. G. Gilliam, "Leukemia mortality. Geographic distribution in the United States for 1949–51," *J. Natl. Cancer Inst.* **17:**475–480 (1956).
47. B. MacMahon, "Geographic variation in leukemia mortality in the United States," *Pub. Health Rep.* **72:**39–46 (1957).
48. A.G. Gilliam and B. MacMahon, "Geographic distribution and trends of leukemia in the United States," *Acta Unio. Int. Contra Cancrum* **16:**1623–1628 (1960).
49. F. Ederer, R.W. Miller, and J. Scotto, "U.S. childhood cancer mortality patterns, 1950–59," *J. Am. Med. Assoc.* **192:**593–600 (1965).

50. T.J. Mason and F.W. McKay, *U.S. Cancer Mortality by County, 1950–69,* DHEW (NIH) Publ. No. 75-780. Washington, D.C.: U.S. Government Printing Office, 1974.
51. T.J. Mason, F.W. McKay, R. Hoover, W. Blot, and J.F. Fraumeni, *Atlas of Cancer Mortality in U.S. Counties, 1950–69,* DHEW (NIH) Publ. No. 76-1204. Washington, D.C.: U.S. Government Printing Office, 1975.
52. F. Burbank, *Patterns in Cancer Mortality in the United States: 1950–67,* National Cancer Institute Monograph, No. 33. Washington, D.C.: U.S. Government Printing Office, 1971.
53. C.R. Stark and A. Oleinick, "Urban or rural residence and histologic type distribution in 21,000 childhood leukemia deaths in the United States, 1950–1959," *J. Natl. Cancer Inst.* **37:**369–379 (1966).
54. A. Blair, J.F. Fraumeni, Jr., and T.J. Mason, "Geographic patterns of leukemia in the United States," *J. Chronic Dis.* **33:**251–260 (1980).
55. S.J. Cutler and J.L. Young, (eds) *Third National Cancer Survey,* National Cancer Institute Monograph No. 41. Washington, D.C.: U.S. Government Printing Office. 1975.
56. S. Selvin, L.I. Levin, D.W. Merrill, and W. Winkelstein, "Selected epidemiologic observations of cell-specific leukemia mortality in the United States, 1969–77," *Am. J. Epidemiol.* **117:**140–152 (1983).
57. G. Wakisaka, H. Uchino, T. Nakamura, S. Shirakawa, A. Adachi, M. Sakurai, and K. Miyamoto, "Present status of leukemia in Japan with special reference to epidemiology and studies on the effect of chemotherapy," *Acta Haematol.* **31:**214–224 (1963).
58. M. Segi and M. Kurihara, *Cancer Mortality for Selected Sites in 24 Countries. No. 6 (1966–67).* Tokyo: Japan Cancer Society, 1972.
59. K. Akazaki and H. Wakasa, "Frequency of lymphoreticular tumors and leukemias in Japan," *J. Natl. Cancer Inst.* **52:**339–343 (1974).
60. A.G. Oettle, "Lymphoreticular malignant neoplasms in South African races. Mortality and morbidity studies," in *The Lymphoreticular Tumors in Africa: A Symposium Organized by the International Union Against Cancer, Paris, 1963,* pp. 1–23. New York: Karger, 1964.
61. J.L. Young, Jr., A.J. Asire, and E.S. Pollack, *SEER Program: Cancer Incidence and Mortality in the United States, 1973–76,* PHS (NIH). Bethesda, Md.: US DHEW 78-1837, 1978.
62. H.F. Dorn, "Illness from cancer in the United States," *Pub. Health Rep.* **59:**33–48, 65–77, 97–115 (1944).
63. P. McPhedran, C.W. Clark, Jr., and J.S. Garcia, "Racial variations in leukemia incidence among the elderly," *J. Natl. Cancer Inst.* **45:**25–28 (1970).

64. H.F. Dorn and S.J. Cutler, *Morbidity from Cancer in the United States: Parts I and II,* US DHEW, Public Health Monograph 56. Washington, D.C.: U.S. Government Printing Office, 1959.
65. M.P. Khokhlova and P.P. Rakhmamin, "Comparative study on geographical distribution of human and cattle leukosis," *Bibl. Haematol.***36:**654–658 (1970).
66. B. Henricson and N. Ringertz, "Comparative studies on the geographical distribution of human and bovine leukemia," *Bibl. Haematol.***30:**331–352 (1968).
67. N.K. Jensen, "A topographical study of leukemia in man and cattle in Denmark," *Bibl. Haematol.* **31:**326–330 (1968).
68. E. Fasal, E.W. Jackson, and M.R. Klauber, "Leukemia and lymphoma mortality and farm residence," *Am. J. Epidemiol.* **87:**267–274 (1968).
69. S. Milham, Jr., *Occupational Mortality in Washington State, 1950–71,* Vol. 1. Contract No. 99074-26, U.S. DHEW, PHS CDC NIOSH, 1976.
70. L. Houten, I.D.J. Bross, and E. Viadana, *A Retrospective Survey of Cancer in Relation to Occupation.* Contract No. HSM99-73-5, US DHEW, PHS CDC NIOSH, 1977.
71. S. Milham, Jr., "Leukemia and multiple myeloma in farmers," *Am. J. Epidemiol.* **94:**307–310 (1971).
72. A. Blair and T.L. Thomas, "Leukemia among Nebraska farmers: a death certificate study," *Am. J. Epidemiol.* **110:**264–273 (1979).
73. K.J. Donham, J.W. Berg, and R.S. Sawin, "Epidemiologic relationships of the bovine population and human leukemia in Iowa," *Am. J. Epidemiol* **112:**80–92 (1980).
74. E. Kvarnfors, B. Henricson, and G. Hugoson, "A statistical study on farm and village level in the possible relations between human leukemia and bovine leukosis," *Acta Vet. Scand.* **16:**163–169 (1975).
75. A. Adelstein and G. White, "Leukaemia 1911–1973: cohort analysis," *Popul. Trends* **3:**9–13 (1976).
76. C.G. Geary, R.T. Benn, and I. Leck, "Incidence of myeloid leukaemia in Lancashire," *Lancet* **2:**549–551 (1979).
77. L. Brandt, P.G. Nilsson, and F. Mitelman, "Trends in incidence of acute leukemia," *Lancet* (Letter to the Editor) **2:**1069 (1979).
78. I.W. Kemp, G.J. Stein, and M.A. Heasman, "Myeloid leukaemia in Scotland," *Lancet* **2:**732–734 (1980).
79. W.R. Gauld, J. Innes, and H.N. Robson, "A survey of 647 cases of leukemia, 1938–1951," *Br. Med. J.* **1:**585–589 (1953).
80. S.I. Kim and C.K. Lee, "Leukemia among Koreans—a statistical and hematological study of 601 cases of leukemia and allied conditions,"

in *Cancer in Asia,* No. 18, pp. 229–237 (T. Hirayama, ed). Baltimore: University Park Press, 1976.

81. M.R. Kushwaha, M. Bagchi, and R.M. Mehrotra, "Leukaemias at Lucknow—a study of 200 cases," *Indian J. Cancer* **15:**28–34 (1978).
82. K. Booth and D. Amato, "Leukaemia in Papua, New Guinea," *Trop. Geogr. Med.* **30:**343–349 (1978).
83. J.C. Bailar, III., H. King, and M.J. Mason, *Cancer Rates and Risks,* Public Health Service Publication No. 1148. Washington, D.C.: US DHEW, 1964.
84. M.H. Griswold et al., *Cancer in Connecticut, 1935–51,* Hartford. Connecticut: State Department of Health, 1955.
85. J.F. Fraumeni, Jr. and R.W. Miller, "Epidemiology of human leukemia: recent observations," *J. Natl. Cancer Inst.* **38:**593–605 (1967).
86. J.F. Fraumeni, Jr. and R.W. Miller, "Leukemia mortality: downturn in rates in the United States," *Science* **155:**1126–1128 (1967).
87. R.A. Kyle, F.T. Nobrega, L.T. Kurland, and L.R. Elveback, "The 30-year trend of leukemia in Olmstead County, Minnesota, 1935 through 1964," *Mayo Clin. Proc.* **43:**342–353 (1968).
88. A. Linos, R.A. Kyle, L.R. Elveback, and L.T. Kurland, "Leukemia in Olmstead County, Minnesota, 1965–1974," *Mayo Clin. Proc.* **53:**714–718 (1978).
89. L. Gordis, M. Szklo, B. Thompson, E. Kaplan, and J.A. Tonascia, "An apparent increase in the incidence of acute nonlymphocytic leukemia in black children," *Cancer* **47:**2763–2768 (1981).
90. M. Szklo, L. Gordis, J. Tonascia, and E. Kaplan, "The changing survivorship of white and black children with leukemia," *Cancer* **42:**59–66 (1978).
91. S.J. Cutler, L. Axtell, and J. Heise, "Ten thousand cases of leukemia: 1940–62," *J. Natl. Cancer Inst.* **39:**993–1026 (1967).
92. C. Zippin, S.J. Cutler, and D. Lum, "Time trends in survival in acute lymphocytic leukemia," *J. Natl. Cancer Inst.* **54:**581–586 (1975).
93. S.J. Cutler, H. Heise, and H. Eisenberg, "Childhood leukemia in Connecticut, 1940–62," *Blood* **30:**1–25 (1967).
94. M.B. Shimkin, E.L. Lucia, K.C. Oppermann, and S.R. Mettier, "Lymphocytic leukemia: an analysis of frequency, distribution and mortality at the University of California Hospital, 1913–47," *Ann. Int. Med.* **39:**1254–1266 (1954).
95. G.R. Minot and R. Isaacs, "Lymphatic leukemia: age incidence, duration and benefit derived from irradiation," *Boston Med. Surg. J.* **191:**1–9 (1924).
96. B.S. Leavell, "Chronic leukemia—a study of the incidence and factors influencing the duration of life," *Am. J. Med. Sci.* **196:**329–340 (1938).

97. C. Zippin, S.J. Cutler, W.J. Reeves, and D. Lum, "Survival in chronic lymphocytic leukemia," *Blood* **42**:367–376 (1973).
98. E.E. Osgood, "Treatment of chronic leukemias," *J. Nucl. Med.* **5**:139–153 (1964).
99. D.R. Boggs, S.A. Sofferman, M.M. Wintrobe, and G.E. Cartwright, "Factors influencing the duration of survival of patients with chronic lymphocytic leukemia," *Am. J. Med.* **40**:243–254 (1966).

4 Personal Characteristics: Age, Sex, Race, and Others

AGE

Age-specific incidence rates from the U.S. Third Cancer Survey demonstrate different age distribution patterns for each of the histopathologic types of leukemias. These different patterns provide additional support for the concept that each histopathologic type may have a different constellation of risk factors [1] (see Figures 3-1 and 3-2 and Table 3-5).

ALL Rates are highest in those under age five (3.8 per 100,000 per year). Incidence rates then decline to a low of 0.2 per 100,000 per year in persons 25–39, after which they then increase to a high of 3.7 in the oldest age groups.

CLL CLL is extremely rare in persons under age 30. Rates then slowly increase from ages 30 to 50. In older age groups, rates exponentially rise to a high of 28.3 in the oldest age group included in the survey (85+).

ANLL Incidence rates are low but stable in persons under age 30 (approximately 0.8 per 100,000 per year). Then rates increase progressively to a high of 14.8 in those ages 75–79.

CML Incidence rates are very low in children and adolescents (0.2 per 100,000) then slowly increase in persons 25–39. Rates then increase more rapidly in the middle-aged.

SEX AND RACE

Sex-specific incidence rates (adjusted by age) from the Third Cancer Survey show higher incidence for males than females among both whites and blacks with all histopathologic types of leukemia, as seen in Figures 3-1, 3-2 and Table 3-5. Age-adjusted rates are also higher in whites than in blacks for all histopathologic types except CML, where incidence rates are virtually the same for whites and blacks.

SOCIOECONOMIC STATUS, EDUCATION, AND INCOME

Data from the Registrar General's Decennial Supplement to Occupational Mortality (1970–72) included information about the relationship between mortality from the leukemias and socioeconomic status in England and Wales. Deaths were classified according to the 8th Revision, but acute and chronic forms of each leukemia subtype were not considered separately. The SMR for males dying of lymphatic leukemia was 115 for Social Class I (the highest class) and 89 for Social Class V (the lowest class). Myeloid leukemia in males exhibited an SMR of 126 in Social Class I and 89 in Social Class V [2].

The prior Decennial Supplement (1961) classified deaths by the 7th Revision, but all leukemias were grouped together. Therefore a comparison of trends by histopathologic type is not possible with this database [3]. Two other large-scale cross-sectional studies that examined the association of socioeconomic variables and leukemia have considered the leukemias as a group; neither found any associations [4, 5].

Devesa analyzed associations between cancer incidence and income and education among whites and blacks for 14 major primary sites (including leukemia) using the population studied in the Third National Cancer Survey [6]. Leukemias were considered by histopathologic type. No associations between ALL, AML, CML, and income or education were found for whites or blacks. Devesa found a negative association in the United States between CLL and income for white males, but none for white females. She also found a negative association between CLL and education for white males but not for white females. The number of CLL cases among blacks in the population she studied was insufficient to identify associations such as those observed among whites. The negative association between CLL and income and education for white males is not consistent with the Registrar General's 1970–72 data in Great Britain. However, the population Devesa studied was completely urban or suburban, whereas the population included in the Registrar General's report was a mix of urban and rural. Also, the method of measuring socioeconomic status was different, both of which may contribute to the disparity in findings.

Comparison of incidence by race, adjusted for age, sex, area, and income or education showed higher rates in whites for all leukemia subtypes except CML [7].

RELIGION AND ETHNICITY

Increased incidence of a disease in a particular ethnic group may suggest directions for exploring etiologic hypotheses. If the occurrence of disease is associated with a particular ethnic group, certain lifestyle factors of that group may be implicated in the etiology of the disease, or in some instances, genetic factors may be important.

Several studies have suggested an excess of leukemia among Jewish populations (see Table 4-1). MacMahon and Koller noted an increased incidence of leukemia in Jews compared to other religious populations among white residents of Brooklyn, 1943–52 [7]. Using the religious affiliation of cemetery of burial as an index for religion and comparing leukemia deaths to a systematic sample of all deaths in white residents of Brooklyn during the same time period, these investigators found that leukemia was listed as a cause of death for Jews twice as often as for non-Jews. The excess risk was found in both sexes, all age groups, both foreign- and native-born groups, and for all cell types; relative risk for CLL was 2.4, AML 2.7, and CML 1.9. Examination of countries of birth showed that a greater percentage of cases were Russian-born, compared to the general population.

Other studies employing the same design and using cemetery of burial and/or the affiliation of the funeral parlor as an index for religion also found an excess incidence in Jews and particularly Jews of Russian and Eastern European origin. These include a study carried out by MacMahon of all cancer deaths among white residents of New York City in 1955 [8], a study of a larger sample of cancer deaths in New York City, 1953–58, carried out by Newill [9], two additional studies of cancer mortality in New York City, 1949–51, among non-Puerto Rican whites analyzed by Seidman [10, 11], a study of cancer mortality in 1950 among foreign-born in 35 states in the U.S. [12], an analysis of leukemia cases included in the Tri-State Leukemia Study (leukemia cases reported from 1959–62 in metropolitan Baltimore, Minneapolis and all metropolitan areas and surrounding counties in New York State) [13], and an analysis of cancer mortality, 1969–71, in Russian-born residents of New York State, excluding New York City, by Greenwald et al. [14]. Two smaller studies of leukemia incidence and religion showed no difference in leukemia rates between Jews and non-Jews [15, 16]. Details and findings of all of these studies are shown in Table 4-1.

Haenszel reviewed several studies of these cancer mortality rates in Jewish populations in order to examine the consistency of their results. These studies include his own report on cancer mortality in the foreign-born in the United States in 1950 with an update for the years 1959–61 [12]; MacMahon's report of cancer mortality in New York City in 1955 [8]; and Seidman's analysis of New York City cancer mortality during the years 1949–51 [10]. Although these studies used different sources of data and methods, Haenszel concluded that their results were consistent. All three studies showed an above-average risk for leukemia in Jews. He also noted that male to female ratios of cancer mortality for several sites, including leukemia, appeared to be closer to one in Jews than in non-Jews [17].

URBAN VERSUS RURAL DIFFERENCES

Comparison of incidence rates of leukemia in rural areas with those of urban areas might suggest etiologic clues if significant differences are found. Data from the Third National Cancer Survey can be used to make such a comparison [1]. Table 4-2 shows age-adjusted incidence rates for males and females: for each of the leukemias, in Iowa (primarily a rural state) and Minneapolis; and overall rates, in the seven cities and two states included in the Third Survey. Minneapolis was selected as a comparison because it is in the midwest, as is Iowa; the population of both areas is primarily white. Iowa has a higher incidence of ALL than does either Minneapolis or all areas combined in the Third Survey. CLL rates in females are higher in Iowa than in all other geographic areas. Rates for males are approximately the same in Iowa and Minneapolis but lower overall for all other areas included in the Third Survey. The difference between Iowa and Minneapolis residents compared to those from other areas is greater for males than for females. CML rates differ very little between these groups.

Within Iowa, there is a definite urban to rural gradient with increased standardized mortality ratios for both ALL and CLL from urban to rural counties [18] (Table 4-3). Standardized incidence ratios were determined separately for males and females with CLL and ALL by urban–rural areas. For CLL, there appears to be a gradient of increasing incidence from urban to rural, particularly for males. There is no apparent trend for either males or females with ALL.

Comparison of urban versus rural incidence rates for each of the leukemias was included in international incidence data published by the International Agency for Research Against Cancer (IARC) in 1976. Age-

Table 4-1 Associations between religion, ethnicity, and the leukemias

Investigator(s) and date	Time period of study	Location of study	Leukemia type(s)	Population studied	Source of ascertainment of religion and/or ethnicity	Findings
MacMahon and Koller 1957 [7]	1943–52	Brooklyn, N.Y.	All types combined Also specific types AML CML CLL	*Leukemia cases selected from:* White residents of Brooklyn which included 1636 patients with leukemia diagnosed 1943–52 from hospital records and death certificates *Controls selected from:* Total white population of Brooklyn A systematic 1 in 200 sample of 1278 deaths from any cause in Brooklyn 1943–52	Religious affiliation of the cemetery of burial was used as an index for religion	Age-standardized annual leukemia (all types combined) incidence rates foreign-born 6.1/100,000, native-born 4.5/100,000 39.6% of the leukemics were born in Russia compared with 26.2% of the total white Brooklyn population Relative risks in Jews compared to non-Jews: AML 2.7 CML 1.9 CLL 2.4
MacMahon 1960 [8]	1955	New York City	All types combined	15,594 deaths in New York City residents in 1955 attributed to neoplastic disease (14,356 deaths were in whites)	*Denominators* Estimates of age and sex distribution were derived by linear interpolation between the censuses of 1950 and 1957 Distribution by nativity estimated by applying to est. 1955 population in each age and sex group the percentage distribution by nativity in the same age and sex group in 1950 Distribution by religion estimated from results of sample survey done by Health Insurance Plan of Greater New York in 1952 *Numerators* Nativity taken from death certificate Religious affiliation based on religious application of cemetery of burial listed on death certificate	Age-standardized annual leukemia (all types combined) incidence rates in whites age 45+: Foreign-born 12.6/100,000 Native-born 7.4/100,000 - - - - - - - - - - - - Jews 19.3/100,000 Catholics 13.3/100,000 Protestants 15.8/100,000

Newill 1961 [9]	1953–58	New York City	Lymphatic Myeloid Acute	*Leukemia cases selected from:* 84,341 deaths of white residents in NYC, 1953–58 *Controls selected from:* Estimates of religious distribution of the population of NYC 2% sample of deaths from all causes in 1955	*Denominators* Computation of rates in 3 main religious groups was based on a sample survey undertaken by the Health Insurance Plan of New York in 1952 *Numerators* Derived from "religious affiliation of cemetery of burial" from death certificates 2% sample of deaths was used to estimate the distribution of all deaths in NYC by religion, nativity, and age	Annual age-adjusted mortality rates per 100,000 by nativity and religion for leukemia by type: Lymph Myeloid Acute Foreign-born 3.0 3.3 3.2 Native-born 1.7 2.6 1.6 - - - - - - - - - - - - Jews 7.0 7.6 3.2 Catholics 4.4 4.5 1.7 Protestants 4.4 5.5 2.0
Haenszel 1961 [12]	1950	35 states in U.S.	All types combined	34,000 cancer deaths recorded for foreign-born whites in 35 states in 1950 (the 35 states accounted for 75% of all cancer deaths among foreign-born whites)	*Denominators* Age-sex distributions for native- and foreign-born whites were obtained from the 1950 census *Numerators* Microfilmed death certificates, listing race and nativity, were obtained from the National Office of Vital Statistics, U.S. Public Health Service	Elevated SMRs for leukemia by sex and nativity: *Males* All foreign-born 107 USSR 145 Czechoslovakia 144 *Females* All foreign-born 112 USSR 182 Poland 157
King, Diamond and Bailar 1965 [15]	1957	Baltimore City and Baltimore County	All types combined	*Denominators* White population all ages residing in Baltimore City and County *Numerators* Death certificates of Baltimore City and County residents	*Denominators* Population estimates of religious preference were determined by interview of a probability sample of whites in the Third Omnibus Survey of the Baltimore market *Numerators* Religious preference determined from funeral home religious affiliation (funeral home obtained from death certificate)	No excess mortality from leukemia in Jewish males or females Age-adjusted mortality ratios *Jews* Males 0.56 Females 0.95

Table 4-1 (continued)

Investigator(s) and date	Time period of study	Location of study	Leukemia type(s)	Population studied	Source of ascertainment of religion and/or ethnicity	Findings
Graham et al. 1970 [13]	1959–62	Tri-State Leukemia Study: All metropolitan areas and surrounding counties in New York State excluding New York City Metropolitan Minneapolis Metropolitan Baltimore	All types combined ALL CLL AML CML unspecified	*Leukemia cases selected from:* All leukemia cases reported to the New York State Cancer Registry and special registries set up in Minnesota and Maryland by the investigators *Controls selected from:* Probability sample of the populations of the metropolitan areas (and surrounding counties)	Personal interview	Age-adjusted relative risks for all leukemias combined: Jews compared with non-Jews 2.44 Foreign-born compared with native born 1.43 Russian-born compared with other foreign-born 2.93 Russian-born Jews compared with native born non-Jews of parents born in foreign countries other than Eastern Europe 5.29 Russian-born non-Jews compared with native born non-Jews with non-Eastern European parents 1.61 - - - - - - - - - - - - Frequency distribution of nationality of adult leukemia cases by histology: USSR compared with Poland, US, and total:

	USSR	Poland	U.S.	Total
ALL	2.8	4.1	8.3	7.7
CLL	52.8	38.9	28.7	31.3
AML	22.2	22.4	22.4	21.9
CML	8.3	22.4	16.9	16.7
Unspec.	13.9	12.2	23.7	22.4

Seidman 1970 [10] Seidman 1971 [11]	1949–51	New York City	All types combined	*Leukemia cases selected from:* Non-Puerto Rican white residents of NYC ages 25–64; deaths ascertained from NYC and New York State death certificates *Controls selected from:* City-wide cancer death rates for the population by sex for a particular cancer site were standardized for age on the total city-wide non-Puerto Rican white population. These rates were then multiplied by the relevant SMRs for each group to obtain average annual death rates per 100,000 population age-adjusted by the indirect method.	*Denominators* Socioeconomic status based on an index constructed from data from the 1950 Census of Population and Housing Religion determined from religious affiliation of cemetery of burial	30% excess leukemia in Jewish males 50% excess leukemia in Jewish females with the differences for males and females persisting across all socioeconomic groupings No association of leukemia with socioeconomic status 40% excess leukemia in Russian-born males 80% excess leukemia in Russian-born females Excess leukemia in both foreign-born Jews and native-born Jews
Greenwald et al. 1975 [14]	1969–71	New York State	All types combined	Russian-born residents of New York State, exclusive of New York City, who died from a neoplasm 1969–71	*Denominators* 1970 census population data for each county in New York State by age, sex, & county of origin SMRs were calculated for selected sites using census file and New York State death certificate file *Numerators* Religion determined by religious affiliation of cemetery or funeral home	SMRs for Russian-born all religions: Males 160 Females 158 Observed/expected Russian-born *Jews* Males 124 Females 108 *Non-Jews* Males 170 Females 118

Table 4-1 (continued)

Investigator(s) and date	Time period of study	Location of study	Leukemia type(s)	Population studied	Source of ascertainment of religion and/or ethnicity	Findings
Cuneo 1976 [14]	1966–74	Nassau County, New York	All types combined	*Denominator* White population all ages of Nassau County *Numerator* All leukemia cases diagnosed in white Nassau County residents, 1966–74	*Denominators* Population figures by religion were based on: National Jewish Population study of 1970–71 which also provided estimates of the age-sex distribution of the Jewish population in Nassau and neighboring Suffolk County An independent estimate was provided by the New York City Council of Churches' Dept. of Church Planning and Research *Numerators* Data on religion was obtained from hospital tumor registries, medical charts, telephone calls to hematologists' offices, death certificate data on cemetery of burial and/or funeral home or from newspaper obituary columns	Mortality rates for Jews compared with Other White by age and sex: *Jews* Males 11.3 Females 8.1 *Other White* Males 12.9 Females 8.6
Haenszel 1971 [17] Review of several sources of cancer mortality in Jewish populations to examine consistency of results including (1) Haenszel (12) (2) MacMahon (8) (3) Seidman (10)			All types combined			

(1) FINDINGS

Male to female ratios of site-specific cancer risks. Jews compared with other population groups

Age-adjusted mortality, NYC 1949–51		Mortality, age 65 and over, U.S. 1959–61		Age-adjusted incidence, Connecticut 1961–65	Age-adjusted incidence, Israel 1960–66		
All Whites	Jews	All foreign-born	Born USSR		All Jews	Oriental Jews	Occidental Jews
1.22	1.05	1.7	1.5	1.6	1.43	1.17	1.6

(2)

Standardized mortality ratios

	Males	Females
U.S. 1950		
Native-born whites	100	100
Foreign-born whites, total	107	112
USSR	145	182
New York City 1949–51		
New York City whites, total	100	100
Jews, all SES classes	132	153
Upper third	142	168
Middle third	119	146
Lower third	133	143

(3)

Mortality rates/100,000 population 65+ (age-adjusted)

	Males			Females		
	U.S.	Metro. counties	Northeast	U.S.	Metro. counties	Northeast
U.S. 1959–61						
Native-born whites	44.8			25.3		
Foreign-born total	44.3	44.0	44.5	26.8	27.2	29.0
USSR	51.0	49.4	50.0	34.5	36.2	40.1

Age-specific rates/100,000

	Males					Females				
	35–44	45–54	55–64	65–74	75–84	35–44	45–54	55–64	65–74	75–84
Native-born whites	3.7	7.0	15.9	36.6	62.7	2.7.	5.0	9.6	20.6	36.1
Foreign-born whites	4.0	7.9	16.8	36.8	62.1	2.7	5.1	11.1	22.0	38.8
USSR	7.2	7.4	22.2	42.5	69.7	5.1	8.5	16.6	24.3	54.3

Table 4-2 Comparison of age-adjusted incidence rates, males and females in Iowa, Minneapolis, and all areas from the Third Cancer Survey

Type of leukemia	Iowa		Minneapolis		Overall rates	
	Males	Females	Males	Females	Males	Females
ALL	2.2	1.0	1.2	0.7	1.5	0.9
CLL	4.0	2.0	4.2	1.5	3.6	1.8
AML	1.9	1.5	2.8	1.7	2.8	1.9
CML	1.8	1.1	1.7	1.4	1.9	1.1

Source: Third National Cancer Survey [1].

adjusted rates for all lymphatic leukemias combined are the same for males in urban areas (3.0 per 100,000) compared with rural areas (3.1), and actually higher for females in urban areas (1.8) compared with rural areas (1.4). Myeloid leukemia (including acute and chronic) is higher in males in urban areas (4.1), than in rural areas (3.3); it is similar in urban females (3.1) and rural females (3.0) [19].

The gradient of increasing incidence of CLL from urban to rural areas of Iowa noted for both males and (less evenly) for females suggests the possibility of an environmental exposure that is most prominent in rural areas but is also found to an increasing extent in semiurban and, even more so, in semirural areas. Possible exposures include pesticides, herbicides, or other agricultural chemicals or other crop or garden-associated exposures; farm animals; "natural" chemicals found in well water, ponds, or lakes. It is interesting that such leukemogenic exposures appear to affect both males and females and therefore are probably environmentally ubiquitous, particularly in rural areas, rather than occupation-specific. The absence of such a gradient in data collected by the IARC may be due to neglecting to perform separate analyses of acute and chronic lymphatic leukemia sub-

Table 4-3 Standardized incidence ratios in urban and rural areas of Iowa by sex for ALL and CLL

Leukemia type	Sex	Urban	Semiurban	Semirural	Rural
ALL	Male	96[a]	77	63	132
	Female	99	122	87	99
CLL	Male	88	97	101	109
	Female	92	83	97	115

[a]Standardized incidence rates (all-Iowa = 100); ratios adjusted for age and time period.
Source: Donham et al. (1980) [18].

types, or the particular leukemogenic exposures occurring in Iowa may not be present in other countries.

REFERENCES

1. S.J. Cutler and J.L. Young, (eds), *Third National Cancer Survey: Incidence Data,* National Cancer Institute Monograph No. 41. Washington, D.C.: U.S. Government Printing Office, 1975.
2. Occupational Mortality Tables, *The Registrar General's Decennial Supplement for England and Wales, 1970–72,* Series DS No. 1, Office of Population Censuses and Surveys. London: Her Majesty's Stationery Office, 1978.
3. Occupational Mortality Tables, *The Registrar General's Decennial Supplement for England and Wales, 1961,* London: Her Majesty's Stationery Office, 1971.
4. J. Clemmesen and A. Nielsen, "The social distribution of cancer in Copenhagen, 1943 to 1947," *Br. J. Cancer* **5:**159–171(1951).
5. S. Graham, M. Levin, and A.M. Lilienfeld, "The socioeconomic distribution of cancer of various sites in Buffalo, N.Y., 1948–52," *Cancer* **13:**180–191 (1960).
6. S.S. Devesa, "A Study of the Association of Cancer Incidence with Income and Education among Whites and Blacks," Ph.D. Thesis. Baltimore: Johns Hopkins University, 1979.
7. B. MacMahon and F.K. Koller, "Ethnic differences in the incidence of leukemia," *Blood* **12:**1–10 (1957).
8. B. MacMahon, "The ethnic distribution of cancer mortality in New York City, 1955," *Acta Unio. Int. Contra Cancrum* **16:**1716–1724 (1960).
9. V.A. Newill, "Distribution of cancer mortality among ethnic subgroups of the white population of New York City, 1953–58," *J. Natl. Cancer Inst.* **26:**405–417 (1961).
10. H. Seidman, "Cancer death rates by site and sex for religious and socioeconomic groups in New York City," *Environ. Res.* **3:**234–250 (1970).
11. H. Seidman, "Cancer mortality in New York City for country-of-birth, religious, and socioeconomic groups," *Environ. Res.* **4:**390–429 (1971).
12. W. Haenszel, "Cancer mortality among the foreign-born in the United States," *J. Natl. Cancer Inst.* **26:**37–132 (1961).
13. S. Graham, R. Gibson, A.M. Lilienfeld, L. Schuman, and M. Levin, "Religion and ethnicity in leukemia," *Am. J. Publ. Health* **60:**266–274 (1970).

14. P. Greenwald, R.F. Korns, P.C. Nasca, and P.E. Wolfgang, "Cancer in United States Jews," *Cancer Res.* **35**:3507–3512 (1975).
15. H. King, E. Diamond, and J. C. Bailar, III, "Cancer mortality and religious preference—a suggested method in research," *Milbank Mem. Fund Q.* **43**:349–358 (1965).
16. J.M. Cuneo, "Leukemia incidence and ethnicity in Nassau County, New York," *Am. J. Publ. Health* **66**:1094–1095 (1976).
17. W. Haenszel, "Cancer mortality among U.S. Jews," *Isr. J. Med. Sci.* **7**:1437–1450 (1971).
18. K.J. Donham, J.W. Berg, and R.S. Sawin, "Epidemiologic relationship of the bovine population and human leukemia in Iowa," *Am. J. Epidemiol.* **112**:80–92 (1980).
19. J. Waterhouse, C. Muir, P. Correa, and J. Powell (eds), *Cancer Incidence in Five Continents, Vol. III,* Sci. Publ. No. 15. Lyon, France: IARC, 1976.

5 Genetic Factors

Several studies suggest that genetic factors play a role in the etiology of some cases of leukemia. A number of different mechanisms appear to be operative. Investigators, using inbred susceptible strains of mice, have shown that the interaction of several genes with other leukemogenic factors (including viruses, chemicals, radiation, etc.) is required in order for most strains of susceptible mice to develop leukemia. The occurrence of familial aggregation of leukemia is somewhat analogous. The genetic substrate required for environmental leukemogens to cause multiple cases in a family is probably polygenic rather than an interaction of environmental factors with a single gene. Subsequent parts of this section discuss other evidence favoring a significant role for genetic factors including data from twin studies, the relationship between abnormal chromosome conditions and the leukemias, the association between human leucocyte antigen (HLA) and the leukemias, and studies of other genetic markers linked with the leukemias.

FAMILIAL AGGREGATION

Many reports of families in which two or more members developed leukemia have been published. Most of the reports indicate that all affected members had the same type of leukemia. In 1977, Gunz analyzed 22 families with 90 affected members. Among the 22 families, 18 (82%) had the same type within each family. In 11 of the 18, all affected members had acute leukemia (not further specified) and in the remaining seven families, all affected individuals had CLL. Familial aggregation appeared to be less frequent for chronic granulocytic leukemia than for ALL, AML, or CLL [1]. Heath reviewed 13 families in which four or more cases of leukemia and lymphoma occurred. In nine of the 13 (69%) families, the affected members within the family had a concordant cell type. For these nine, the frequency of histopathologic type included two with CLL, two with ALL, two with AML, and three with acute leukemia, cell type not further specified [2].

Closer examination of some of the families suggests some clues as to the mechanism by which leukemia occurs. Heath pointed out evidence of consanguinity in at least three families [2]. A study by Kurita and Kanei in Japan supported the hypothesis that inbreeding can increase the risk of leukemia. They noted that consanguinity occurs more frequently when two or more sibs develop leukemia than when two or more cases are less closely related than siblings [3].

Cervenka carried out chromosome analysis in five families with two or more leukemia cases. In one of these families with two cases of leukemia (both ALL), the investigators found that a large proportion of the unaffected first-degree relatives had a chromosome abnormality suggesting a possible association between the high frequency of chromatid exchange figures and the excess occurrence of leukemia [4].

In another study, immunologic function was examined in parents of children with acute leukemia and compared with parents of controls. Mothers of cases were found to have lower absolute monocyte counts, higher levels of gamma globulin, IgA, and IgG than control mothers. Three possible explanations were proposed by the investigators. The immunologic factors identified represent a familial immunologic abnormality, or mothers of leukemia children may be carriers of an infectious agent, which when passed to one or more susceptible children may lead to the development of leukemia. These infected mothers do not have clinical manifestations of the infection. A third hypothesis is that mothers of leukemics are exposed to infectious agents more frequently than are mothers whose children do not have leukemia [5]. A study of one family in which CLL developed in four of five siblings whose father also had CLL showed that, in three of the siblings, peripheral blood leukemic cells shared the same type of heavy and light chains as the only detectable surface immunoglobulin. This suggested that, at a cellular and molecular level, the CLL in family members was identical. The conclusion was that an inherited defect in the class of cells destined to express that type of heavy and light chain as a surface immunoglobulin was the underlying cause for increased susceptibility to leukemia in the family [6].

Recently, Gunz et al. reported 14 persons with leukemia in three generations of a family consisting of 292 members. Thirteen of the 14 cases had acute leukemia (nine with AML and four unknown because of absence of medical charts or complete diagnostic evaluation); one family member developed CML. HLA typing, chromosome analysis, and study of 31 serum protein and red cell enzyme polymorphisms were unremarkable. HLA types and haplotypes were not associated with leukemia. Detailed immunologic studies could not be carried out in the family. No autoimmune or immunodeficiency diseases were known to have occurred in the family. Intracytoplasmic particles similar in size and density to a C-type virus were found in blood cells of two leukemics in this family; biologic activity and budding of virus particles were not detected. Interviews with family members suggested no significant exposure to radiation, benzene, leukemogenic drugs, or animals with high incidence of leukemia (including chickens, cats, and cattle). However, many members did have contact with birds (budgerigars), tropical fish, and dogs, particularly greyhounds [7].

Details of additional reports of familial aggregation of leukemia can be found in a number of reviews [1, 2, 9] as well as in the original papers [10–31].

In general, investigators have not identified blood cell or other markers that identify family members who develop leukemia in multiply affected families. Goudsmit et al. reported a family in which five siblings had an unusual finding, Döhle inclusions in the granulocytes; two of the five developed AML [8].

Multiple cases of leukemia could occur by chance in isolated families, but the occurrence of concordant types of leukemia in the majority of affected individual families mitigates against this conclusion. Individual family reports are difficult to interpret since such families may be highly selected and not representative of all families with two or more members with leukemia. Families with cases of a concordant type might be more likely to be reported than others with different types within a family. Isolated family reports do not permit one to estimate the expected numbers of the leukemias and do not allow determination of whether two or more cases in a family represent an excess. To determine whether incidence of leukemia is increased in families, families of index leukemia cases must be compared with families of index controls.

Vidabaek studied 209 persons with all types of leukemia and found 17 (8.1%) with a family history of leukemia compared with one (0.5%) with a history among 200 controls. He also noted an increased incidence of cancer in case families compared with controls [32]. Vidabaek's findings have been criticized. Cases were not identified by a population-based approach but were a highly selected group identified from hospitals in greater Copenhagen. Only 209 of 310 cases located were included. Virtually all of the patients included in his study had died and information was obtained from medical charts and next-of-kin. Based on cancer statistics in Denmark, the families of leukemia cases had normal cancer incidence rates whereas the control families had a lower cancer incidence rate than normal. Affected family members were not specifically defined for the purposes of the study. Very distant as well as closer relatives were included among the 17 relatives reported to have leukemia in families of cases. Families with one member with leukemia are probably more likely to be aware of leukemia in distant relatives than are those in control families. All types of leukemia were included among the 209 cases. Similar studies have compared the frequency of leukemia and/or cancer in general in families of patients with leukemia with families of controls or with vital statistics data and found no significant difference between case families and controls [17, 33, 34]. Most of these studies had similar methodologic problems as the one by Vidabaek: cases were a select group from a single hospital or may have been somewhat more representative but incompletely followed-up; case-group sample size was small; controls and their relatives were incompletely described or not as carefully followed-up as cases and their relatives; again, affected family members included both very distant as well as closer relatives instead of a more conservatively defined (and easier to follow-up) group such as first-degree relatives.

However, other studies have noted an excess of leukemia and/or cancer in general in families of leukemia cases compared with controls [24, 35–37]. In one study carried out by Gunz et al., leukemia patients including all cell types were used as the case group [36] and in another study, CLL cases alone comprised the case group [24]; in yet a third study, all cell types were included but analyzed separately by cell type [37]. Matched controls were used in the first study; population incidence rates were compared to case-family leukemia incidence in the second and third surveys. For the family study employing all leukemias combined as index cases (590 index patients and their families), 36 families (6.1%) were found to have additional leukemia cases compared with 17 families (2.4%) among 712 controls [36]. Comparison of incidence of leukemia in first-degree relatives of CLL cases with population leukemia incidence rates showed 7 times more leukemia cases in relatives of CLL patients than expected [24]. From the Third Survey, leukemia as a whole was again found to have increased familial incidence; for first-degree relatives, the ratio of observed to expected was 2.42. The proportion of leukemia cases by cell type with first-degree relatives with leukemia was 2.0% for acute leukemia (not further specified by histopathologic type), 3.5% for CLL, and 0.8% for CML. The familial incidence was most marked for CLL (which constitutes 30% of leukemias in western nations), less so in acute leukemias (50% of all leukemia cases), and probably not in excess for CML (comprising 20% of leukemias) [37].

Should the apparent familial excess found in several studies in Australia and New Zealand by Gunz and his colleagues be confirmed in additional cell type-specific studies, genetic factors as well as shared environmental leukemogens must be considered when attempting to explain the finding. The specific nature of the genetic factors, environmental factors, or their interaction resulting in excess familial leukemia incidence deserves additional investigation. Epidemiologic, population-based family studies, including meaningful clinical and laboratory components, systematically carried out on families with two or more cases of leukemia and appropriate control families might be very informative. For example, immunologic studies using current knowledge and techniques could provide clues as to possible causes for the occurrence of two or more cases in a given family.

TWIN STUDIES

Another type of study that may be helpful in investigating the role of genetic factors is the examination of leukemia occurrence in twins carried out in a methodologically rigorous fashion. The rationale for using twin studies to delineate the potential etiologic role of genetic factors is hypothesized somewhat as follows. Abnormalities of, or mutations involving, chromosomes and/or single or multiple genes may occur in the parents' fetal, childhood, or adult existence prior to conception of their affected offspring. Damage to the genetic material of the affected twins may also take place during the earliest divisions of the zygotes. Such abnormalities involving genetic material would therefore be more likely to result in higher concordance of diseases such as leukemia (assuming genetic factors were etiologically important) in monozygous compared to dizygous twins. However, similarly, leukemogens acting upon genetic material of dizygous twins would, most likely, also be exerting toxic effects during the earliest divisions of the zygotes. However, should one find that twins are discordantly affected, i.e., only one develops leukemia, then theoretically environmental factors are more likely to be responsible.

There are a number of problems with this hypothesis. Environmental factors such as viruses, drugs, or toxic chemicals may be transmitted to the monozygous twin embryos through the placental cross circulation early in fetal development without causing damage to genetic material and yet subsequently producing leukemia. No twin studies have been reported that demonstrated concordance rates of 100% for leukemia. It has been postulated that a "significant" rate of concordance, arbitrarily defined as 20%–25%, has important implications with respect to the importance of prenatal factors in the etiology of leukemia. However, the arbitrariness of the "significant" levels of concordance chosen to implicate prenatal or genetic hypotheses must be considered. In fact, one must try to answer the related question of why concordance is not 100% if genetic factors are etiologically important as initiating factors. MacMahon has suggested that there may be lack of concordance for postnatal *promoting* factors [38]. Another important methodologic issue is the laboratory and other methods used to determine zygosity. Many tests defining a number of genetic markers have been reliable only in the last 10–15 years; therefore, study conclusions based on zygosity published prior to this time must be interpreted cautiously. Finally, case reports of leukemia concordance or analyses of data not carefully validated in terms of zygosity status and diagnosis of leukemia may be subject to selection biases. The types of approach that ought to be employed are (1) use of large twin registries employing population-based high levels of twin reporting; and (2) national or even international population-based studies of twin pairs.

Several case studies of leukemia occurrence in twins have been reported [39–45]. Zuelzer and Cox noted that, in 25 reported cases for which adequate detail was available, the majority of concordant leukemia cases in twins were acute leukemia (type not further specified). Young infants predominated in the reported cases of concordant acute leukemia in monozygous twins. One of the 25 twin pairs had CML and three were reported to have CLL. Only a few cases of discordance for CML in monozygous twins have been reported. Zuelzer and Cox noted that it was more likely that concordant cases would be reported rather than discordant cases [46].

Several studies of large series of leukemia cases have determined the frequency of concordant leukemia twin pairs in such case groups. Court-Brown and Doll found only one possible case of concordant leukemia in identical twins in a retrospective study of 5,143 children dying of leukemia among whom 20 monozygous pairs of twins were expected to occur. One member of the pair was certified as dying of leukemia while the cause of death of the other member was given as aplastic anemia. The investigators did not state the actual number of twin pairs occurring in this group of children [47].

MacMahon and Levy found three twin sets with concordant leukemia in a series of 3,472 childhood leukemia deaths obtained from the National Center for Health Statistics for the northeastern United States. The database was augmented by three smaller groups of cases from individual medical centers, increasing the total number of children with leukemia to 4,679, among whom there were 72 twin pairs. The number of sets found to be concordant for leukemia was five. While only two of the five pairs in which both twins were affected had been studied in detail, the investigators believed that available evidence (based on blood typing, statements on the birth certificates, age of mothers at birth of the twins, and similarity of these five pairs in age of onset and death from leukemia to other monozygous pairs reported in the literature) suggested that all five pairs were monozygous. Since none of the definite 25 unlike-sex twins were concordant for leukemia (and assuming that there were equal numbers, or 25 pairs, of like-sex dizygous twins as there were unlike-sex twins), they estimated that there were 22 pairs (72 − 50) of monozygous twins with a concordance rate of about 25% (5/22) [38]. This estimate, however, must be considered as having wide confidence intervals as well as being somewhat speculative since it is based on a number of assumptions.

Miller analyzed death certificates for 21,659 children under age 15 who died of cancer in the U.S., 1960–64. Among the cancer deaths, 10,390 were due to leukemia. Five pairs of like-sex twins with leukemia, all under age six at death, were identified. The exact number of twins in the series was unknown but was estimated to be 156. Since the number of twin pairs at risk according to sex and zygosity was not known, Miller was not able to estimate the concordance rates. No sets of twins of unlike sex were concordant for leukemia and no sets of twins died of other forms of cancer although the group dying of other cancers included 11,269 children with an estimated 170 twins. Five pairs of nontwin sibs died from leukemia [48]. An extension of this five-year nationwide survey of death certificates for another three years yielded two additional identical sets of twins dying of leukemia under age seven. With this additional data, the concordance rate was estimated to be 17%. Examination of the leukemias by cell type indicated apparent concordance by cell type, although 8 of the 14 cases were identified only as acute leukemia [49].

In their review and analysis of reported concordant cases of leukemia in monozygous and dizygous twins, Zuelzer and Cox noted that the age of onset of leukemia was under one year in 10 of the 16 pairs believed to be monozygous. The distribution in the five pairs of dizygous twins was closer to that of nontwin children with an age peak between three to five years of age. The male to female sex ratio was 7:9 for the monozygous whereas it was 4:0 (excluding the one unlike-sex pair) for the dizygous pairs [46]. In view of the age distribution of onset and the concordant occurrence of leukemia in monozygotic twins, which is skewed to infants under one year of age, the chance of onset of leukemia in the second member of a pair of monozygous twins in which the first developed the neoplasm under age one may be higher than the 25% concordance estimated by MacMahon and Levy. The risk in dizygotic twins may be much lower. Though this theory may be consistent with an etiological role of genetic factors, it may also be indicative of other causes. Among possible explanations, one must consider the fact that some twins share a common intrauterine environment and monozygotic twins share a common circulation, and it is also possible that exposure to *in utero* leukemogens could likely affect both members of a pair.

In 1963, Pearson et al. concluded that, although the true incidence of concordant leukemia in identical twins was greater than could be explained by chance, it was less than expected if genetic factors played a major role [45]. Later data from the large series described above appear to support this conclusion.

Although the number of twin pairs concordant for leukemia is very small in spite of the large numbers of leukemia cases from which the twin pairs were identified, it is unlikely that larger studies would be feasible.

STUDIES OF SPECIAL POPULATIONS

In addition to familial aggregation, the examination of leukemia occurrences in genetically isolated, inbred populations is another way to study the role of genetic factors. Feldman et al. found an excess of ALL in young females in an isolated Syrian Jewish community in Brooklyn that was 30 times greater than expected. Risk factors were believed to include a high frequency of consanguinity (the frequency of first-cousin marriages was estimated to be 7%) and the older age of fathers compared to community and noncommunity sets of controls [50].

CHROMOSOMAL ABNORMALITIES

Historical Background

The role that genetic factors play in the development and natural history of all of the leukemias has been redefined and expanded since 1960 when Nowell and Hungerford described the association of the Philadelphia or Ph^1 chromosome (thought to arise from a simple deletion with translocation of the deleted long arm of a G chromosome) with chronic myelogenous leukemia [51]. A major breakthrough in the study of chromosomes occurred in 1969 when Caspersson et al. first reported a method for distinguishing chromosomes that were quite similar. The method, called banding, was based on a differential binding of fluorescent derivatives of quinacrine to different parts of chromosomes [52–54]. Following the development of banding techniques, a number of studies showed that 50% of patients with acute leukemia were found to have chromosomal changes in leukemic cells, although the exact percentage appeared to vary between laboratories from 30% to 65%+. In the last few years, new high resolution techniques have been developed suggesting that a much higher proportion of persons with acute leukemia have chromosome abnormalities. Recently (1981) Yunis et al. reported a high resolution methotrexate cell-synchronization technique for chromosome analysis in which he and his co-workers found abnormalities in all of his 24 study patients with ANLL [55]. Chromosomal abnormalities associated with the leukemias are almost completely confined to the leukemic cells. Subtypes of the various leukemias characterized by the specific chromosome abnormalities have been identified. The subtypes with different abnormalities often have distinctive clinical and laboratory features, as well as different responses to treatment.

The various findings about the correlation between presence of chromosomal abnormalities and prognosis have also been noted for preleukemia. Some investigators have found that chromosome abnormalities in preleukemia have a poor prognosis [56–58]. Others have failed to show such a correlation [59]. Still others have had mixed findings with respect to the association [60].

Methodologic Issues

Several methodologic problems must be considered when attempting to interpret the significance of various chromosomal abnormalities associated with each of the leukemias. The temporal relationship between the appearance of an abnormality and onset of leukemia is poorly understood. To elucidate this relationship, it is necessary to do serial chromosome analyses preferably using the most advanced technology (i.e., high resolution techniques) in large populations at high risk for leukemia, such as workers exposed to high levels of benzene. Clearly such studies are not now feasible. However, case-control studies might be able to provide indirect evidence about whether the role of the chromosome abnormality is of a causal nature if population-based, rather than highly selected, cases are identified and the diagnostic work-up and chromosomal analysis is performed prior to progression of disease and administration of treatment. This is particularly essential if epidemiologic data is correlated with particular chromosome defects.

The exact temporal sequence is also difficult to determine because many of the suspected risk factors for leukemia, such as infectious agents (viral hepatitis, varicella), drugs (antineoplastic agents), chemicals (benzene), and ionizing radiation have been repeatedly shown to cause chromosomal abnormalities in animal studies as well as *in vitro*. Hence, it may not be particularly important to determine the sequence of events.

Given the rapid advances in the technologies used for chromosome analysis in recent years, a larger proportion of persons with ALL and ANLL have been found to have chromosomal abnormalities. Little correlation between chromosomal abnormalities and epidemiologic factors has been found.

Other than a number of studies of Down's syndrome patients there have been virtually no systematic studies of groups with other chromosomal abnormalities (Fanconi's anemia, Klinefelter's syndrome, etc.) who are at high risk of leukemia. Careful epidemiologic and chromosomal study of such high risk groups and their families may lead to important insights about interactions between the abnormal chromosome features and environmental factors that may result in leukemia. Clearly the rarity of these conditions would make it necessary to carry out a national study or may even require an internationally coordinated effort.

The newest technology involving high resolution techniques for chromosomal analysis is very time consuming and available only in a few laboratories. Until such technology becomes more widely available and less expensive, large-scale epidemiologic studies that include chromosomal analysis will not really be feasible.

Chromosomal Abnormalities in the Leukemias

ALL Abnormalities noted in studies of ALL cases prior to the development of banding techniques include: some cases with hypodiploidy (less than the diploid number of chromosomes; i.e., less than 46); the occurrence of modes, defined as the predominant number of chromosomes in tumor tissue whose cells have a varying number of chromosomes, near and beyond the tetraploid range, i.e., twice the number of chromosomes in diploid cells; some instances of morphologically ill-defined and fuzzy-appearing chromatids; and frequent failure of the chromosomes to show the characteristic banding pattern of normal chromosomes [54, 61–64]. In various series, between 50% and 79% of ALL patients have been shown to have aneuploid clones, that is, a group of cells believed to derive from a single abnormal cell, that have a number of chromosomes that varies from the normal 46. The distribution of aneuploidy and diploidy is similar in untreated and treated groups of patients suggesting that the chromosomal abnormalities are present prior to therapy. However, following successful therapy, the aneuploid metaphases may completely disappear, whereas the karyotypically (karyotype designates the chromosome complement) abnormal cells may reappear when relapse occurs [52]. Therefore, cytogenetic findings may be useful in evaluating disease status, particularly in ALL.

Chromosomal changes in ALL frequently involve C- and G-group chromosomes. The abnormalities are probably distributed nonrandomly, i.e., the long arm of #6 is frequently deleted [65, 66]. Other abnormalities reported include the presence of a Ph^1, a missing Y, and various translocations. In most cases, the abnormal clone of cells (often characterized by one or more specific chromosomal abnormalities) noted at diagnosis, which disappears during successful treatment and remission, is the same one that reappears during relapse. Occasionally, the relapsed clone is believed to have an independent origin; its histopathologic and chromosomal abnormalities may be quite different than those of the original abnormal clone.

Several investigators have reported karyotypic instability, selection and/or evolution during the course of ALL [64, 66, 67]. No clear-cut prognostic and clinical differences have been established for the various aneuploid and diploid forms of ALL cell abnormality. Efforts are now underway to correlate cell surface markers, other clinical, laboratory, and survival parameters with chromosomal abnormalities in ALL.

Certain congenital chromosomal abnormalities have been reported to be associated with ALL. Diseases with these abnormalities include males with Klinefelter's syndrome (XXY or XXXY), a child with XYY, and several children in the same family with various translocations [68–70].

Rare cases of ALL are Ph^1-positive. With banding techniques, the Ph^1 in ALL has been shown to be a deleted chromosome #22. Sandberg [52] summarized the eleven published case reports of this chromosome abnormality in ALL. In eight of the eleven patients in whom a complete remission was achieved, a normal marrow karyotype was noted. Each of the eight patients had some normal cells present in the marrow before therapy.

CLL Knowledge of chromosomal abnormalities in CLL is very scanty primarily because of technical problems of chromosome preparation related to the peculiar pathophysiology of the disease. CLL is probably a B-cell malignancy characterized by an accumulation of small lymphocytes, a slow cellular proliferation and dividing rate of cells, and a poor response of the leukemic B-lymphocytes to various mitogens. Currently available mitogens act upon T-cell lymphocytes and, only slowly and after a delay, upon leukemic CLL lymphocytes. Reports of chromosome analysis done on CLL patients must be interpreted cautiously, especially if the cells have been incubated with the mitogen for less than five days [71–76].

The majority of cell metaphases of patients with CLL that have been examined have been normal. However, this may be because observations are primarily made on T-lymphocytes, instead of the leukemic B-lymphocytes, as noted above. Some of the few abnormal cases reported have: a small number of hyperdiploid metaphases (47–48 chromosomes), pseudodiploidy with no consistent anomalies seen in the studied cells; a significant percentage of aneuploid cells (aneuploidy mainly characterized by hypodiploidy) [74, 77].

An interesting abnormality, the Christchurch (Ch^1) anomaly, was found in four sibs in a sibship of seven [78, 79]. Three of the four carriers developed CLL. Fitzgerald et al. implied that Ch^1 might cause the carriers to be at higher risk for CLL. Ch^1, however, is quite different than Ph^1

because it is an inherited abnormality present in all tissues, not only the leukemic cells. However, Ch^1 has not been found in other instances of CLL nor has CLL been found in families with Ch^1 [25, 80–82].

Recently a rapidly progressive case of CLL was observed in a 10-year-old girl, whose leukemic cells bore surface IgM and IgD immunoglobulins of the kappa-light-chain type. When the abnormal cells were karyotyped they showed an unusual translocation [t(2;14) (p13;q32)] with break points at or near the kappa-light-chain and heavy-chain gene loci. These translocations involving the immunoglobulin gene loci may, analogous to the situation with Burkitt's lymphoma, possibly be important in the pathogenesis of some cases of CLL. The correspondence between the involved light-chain gene and the monoclonal immunoglobulin light-chain type expressed on the neoplastic cell surface has also been reported in Burkitt's lymphoma [83–85].

ANLL Acute nonlymphocytic leukemia cases may have hypo-, pseudo-(46 chromosomes but an abnormal karyotype), or hyperdiploidy, but seldom a chromosome number exceeding 55. Occasionally very low chromosome numbers may be found. The percentage of aneuploid cells may also vary from 10% or less to 100%. The reappearance of cells with abnormal numbers or types of chromosomes in persons with ANLL usually is associated with imminent relapse [52].

Similar to the situation with ALL, there have been reports of ANLL cases with initially karyotypically normal cells which developed cytogenetic abnormalities or have shown a definite progression of the baseline chromosome abnormalities. Many Ph^1-positive cases of ANLL have been described. Most cases studied with banding have been shown to have the standard Ph^1 translocation but several with unusual translocations have also been described [52].

Chromosomal changes noted in ANLL most commonly involve chromosomes #7, #8, and #21. Ten percent to 15% of patients with ANLL have a reciprocal translocation between #8 and #21, often accompanied by the loss of a sex chromosome, particularly in males [86–88].

Certain cytogenetic abnormalities occurring in ANLL may be used to classify the disease as well as determine prognosis. Sakurai and Sandberg [86] noted that AML patients, the most carefully studied subgroup of ANLL cases, with no normal karyotypes in their marrow (about 15% of AML cases) had a shorter survival than that of patients with a mixture of normal and abnormal karyotypes or patients with completely normal karyotypes. They postulated that, if patients with no normal karyotypes are treated aggressively with chemotherapy, when the leukemic cells totally disappear from the marrow, no normal elements are available for repopulating the marrow. To further support this theory, the authors compared the distributions of chromosome abnormalities in AML with those in ALL. With rare exceptions, all of the ALL patients examined cytogenetically have had either completely normal karyotypes or a mixture of normal and abnormal karyotypes compared with the 15% of AML patients that have no normal karyotypes. This finding may explain, to some extent, the more favorable prognosis of ALL patients compared to those with AML. Other workers have found very little association between karyotype and AML survival [89]. At the First International Workshop on Chromosomes and Leukemia, van den Berghe noted a correlation between cytogenetic patterns of marrow cells and survival for AML but not in patients with myelomonocytic or monocytic varieties of ANLL [90]. In view of the other factors known to be associated with decreased survival (increased age, etc.), the contribution of each factor toward decreased survival should be examined by means of multivariate regression analysis.

Examination of chromosomal abnormalities in adults with ANLL has shown certain characteristic patterns. These include an additional #8 chromosome, a loss of a #7 chromosome, either a gain or a loss of a #21 chromosome, and a translocation involving chromosomes #8 and #21 with or without the associated loss of a sex chromosome. A 15:17 translocation in acute promyelocytic leukemia has been noted [87, 91–93]. Some investigators have shown that adult patients with ANLL have a poor prognosis if their bone marrow cells are cytogenetically abnormal, especially if no normal metaphases are found [94–96].

Abnormal cytogenetic patterns have also been found in 50% of children with ANLL. A variety of chromosomal changes have been found. Children with ANLL who have chromosomal abnormalities in their bone marrow sample had a significantly shorter median survival than those with a normal diploid pattern [97].

Acute promyelocytic leukemia (APL) is a rare form of ANLL; it constitutes about 5% of all acute leukemias in adults [98]. A translocation between the long arms of chromosomes #15 and #17 was first reported in 1978 in 9 of 17 cases of APL but not in any other subgroup of ANLL [99]. Van den Berghe et al. reported that 14 of 16 Belgian patients with APL had the same translocation [100], but Teerenhovi et al. in Finland and Sweden found that only 1 of the 12 patients with APL that they examined cytogenetically had the translocation [101].

Many issues remain unresolved with respect to the association of chromosome abnormalities and ANLL. First, it is unclear whether chromosomal changes result from the factors causing ANLL or whether these changes are the cause of the leukemic process. In a significant number of cases, some evidence shows progression from diploidy to aneuploidy following diagnosis of ANLL [102–104]. Banding techniques have also shown that the evolutionary process of the disease may involve acquisition of an additional #8 chromosome (similar to the addition of a #8 in CML when blastic transformation occurs) or the loss of a #7 (also similar to CML in blastic transformation) [87]. A second issue is whether or not ANLL is a clonal disease, in view of the karyotypic instability that has been reported [87, 105]. A third unresolved issue is whether or not ANLL, causally associated with benzene poisoning, radiation, or a drug, has distinctive cytogenetic patterns specifically caused by the particular exposure or agent.

Congenital leukemia is very rare; AML is the predominant form when it does occur and also when a leukemia develops in the first year of life. Some congenital leukemias are cytogenetically normal; others show a variety of karyotypic changes [52]. Sandberg et al. observed hypodiploidy in one and hyperdiploidy in another of a pair of fraternal twins with congenital AML [106]. Any implication from this should be made cautiously but it is possible that the same causal factors may result in different cytogenetic forms of AML and that different cytogenetic forms need not be classified as different subtypes of the disease.

The incidence of leukemia, especially AML and ALL, appears to be increased in individuals with Down's syndrome when compared to the general population [107–110]. Families with a high incidence of leukemia appear to have a high incidence of Down's syndrome though the two diseases do not always involve the same individuals [111–113]. In a review of the 276 recorded cases of Down's syndrome and leukemia in the world's literature, Rosner and Lee found that, among the 47 newborns in this group who developed leukemia, 58% had AML and 42% had ALL, whereas, among the remaining 229 children with Down's syndrome who developed leukemia, 31% had AML and 69% had ALL [114]. The chromosomal changes in AML and ALL patients with Down's syndrome parallel the chromosomal changes in leukemics without Down's. The characteristic chromosome abnormality associated with Down's syndrome is trisomy #21. Other than advanced maternal age, causal factors of trisomy #21 are largely unknown. A second, much less common type of Down's syndrome is due to translocation, either G–G or G–D [115].

CML Chronic myelogenous leukemia is the only human malignancy with a consistent karyotypic abnormality, the Philadelphia chromosome (Ph^1). The Ph^1 chromosome is defined as a postzygotic anomaly most often associated with myeloproliferative diseases, predominantly CML. The evidence supporting the postzygotic, acquired nature of Ph^1 is: (1) marrow cells of first degree relatives of affected individuals have not been shown to have Ph^1 [116]; (2) twin studies of CML found that the hematopoietic cells were Ph^1 positive in the CML-affected twin only [117–120]; (3) CML patients do not have the Ph^1 in nonhematopoietic cells [121, 122]. The development of autoradiographic techniques followed by the use of the fluorescence banding technique and the Giemsa banding technique have made possible the identification of the origin of the Ph^1 and the translocation of the material. Measurement of DNA by means of quantitative cytochemistry, scanning cytophotometry, and image-processing techniques has revealed a balanced reciprocal translocation between #22 and #9. This balanced translocation is significant because it may be that no DNA is actually lost from the cell. However, current methods will not detect loss of very small amounts of genetically critical DNA.

Other Ph^1 translocations have been reported in less than 10% of cases with CML including 2:22, 3:22, 6:22, 10:22, and a number of others. In addition, a number of "complex" translocations have been described which are usually between #9 and #22 but different than the much more common balanced reciprocal translocation. The clinical, hematologic, and prognostic characteristics of these cases do not appear to be significantly different from those of CML cases with the more standard Ph^1 translocation [123].

The mechanism by which the Ph^1 chromosome is acquired is unknown. It is also unclear if the chromosomal change is causally involved in the development of the disease. To date, studies have not clarified whether there are genetically predisposed individuals who are susceptible to breakage in the q11 region of chromosome #22, whether the two-hit model applies (i.e., the first "hit" makes the individual metabolically, hormonally, etc., susceptible to breaks which then occur after the second "hit" of radiation, drugs, etc.), or whether a single factor simultaneously causes both the chromosome break and CML.

Most evidence favors the clonal or single cell origin and evolution of CML [124–128]. For the most part, there is good agreement that, in the majority of cases of CML, the Ph^1 originates in an early precursor cell which can differentiate into myelocytic, erythrocytic, or megakaryocytic cells. Clonal evolution is defined as a stepwise rearrangement of the karyotype proceeding in an orderly way. One piece of evidence is based on cytogenetic and enzymatic studies in females (with Ph^1-positive CML) of the sex-linked enzyme glucose-6-phosphate dehydrogenase (G-6PD). Both granulocytes and erythrocytes of these females with CML who were heterozygous for this enzyme were found to carry only one enzyme allele governing production of G-6PD. Other evidence is the finding that the Ph^1 abnormality occurs in myelocytic, erythrocytic, and megakaryocytic cells. However, somewhat contrary to the clonal evolution theory is evidence suggesting that the Ph^1 chromosome originates in two or more precursor cells [129, 130].

In addition to the two major subgroups of CML, those who are Ph^1-positive (85%–90%) and those who are Ph^1-negative (10%–15%), there appear to be additional subtypes of Ph^1-positive disease: Ph^1-positive CML with missing Y, Ph^1-positive cases with additional karyotypic abnormalities, and Ph^1-positive CML with subpopulations of Ph^1-negative hematopoietic cells. Clinical differences appear to exist between the karyotypically different types. These are shown in Table 5-1 [131–134].

Approximately 8%–10% of males with CML have been described as having only 45 chromosomes in the marrow cells with the Y chromosome missing [135]. The loss of the Y is restricted to hematopoietic cells and appears to be secondary to the appearance of the Ph^1 chromosome in this group; a case has been reported in which loss of the Y occurred as the disease progressed [136]. While the missing Y may be a normal aging phenomenon (noted in approximately 10% of normal males over the age of 60), it may be causally related to CML and may also have prognostic significance [52]. The mean age of males who have Ph^1-positive CML and the missing Y is 49. The median survival as can be seen in Table 5-l is 48 months, longer than for other subtypes, and the disease can be benign in these males. The loss of the Y chromosome is also seen in other hematopoietic diseases including polycythemia vera, megaloblastic anemia, and chronic myelomonocytic leukemia. The missing Y or XO marrow has been noted to disappear with therapy in a proportion of cases [52]. The XO pattern has not been reported in females.

It is possible to correlate clonal evolution and the natural history of CML. The first and usually only initial karyotypic abnormality in early CML is Ph^1-positivity. Aneuploidy is often a later manifestation of CML, tending to occur at least two years after diagnosis. Once aneuploidy develops, prognosis is usually worse. Chemotherapy may lead to repopulation of the hematopoietic cell subpopulations that are chromosomally normal. Gradually, Ph^1-positive cells return and, then, cytogenetic progression occurs. In the late stages of CML, especially in the blastic phase, normal cells are very rare and striking chromosomal changes of the leukemia cells are usually present. Even if patients with Ph^1-positive CML who are in the blastic phase go into remission, Ph^1-positive cells do not disappear from the marrow. Accompanying the cytogenetic progression in the blastic phase is an increase in the proportion of myeloblasts, which correlates with a worsening of prognosis [52].

A few cytogenetic studies of CML cases have been carried out in black African populations. Leibowitz found the Philadelphia (Ph^1) chromosome in 19 of 20 black South Africans with CML [138]. Median age of onset was 45 years with a range of 27–65. Males predominated by a ratio of 2.5:1. Several unusual clinical features were present in the 20 cases: significant lymphadenopathy, congestive cardiac failure secondary to leukemic infiltrates, and skin lesions in a significant number of cases in addition to the more common splenomegaly. Another uncommon clinical

Table 5-1 Clinical differences between karyotypically different subtypes of CML

CML karyotype subtype	Sex	Leukocyte count	Platelet count	Hct	Spleen	Peripheral smear	Response to therapy	Survival	Other
Ph^1-positive (100% of metaphases)	Equally M & F	100,000/mm^3	Normal or elevated	Slight anemia	Markedly enlarged	Only occasional myeloblasts	Good to excellent	Median is 40 months	—
Ph^1-positive (100% of metaphases) and missing Y	Males only	—	—	—	—	—	—	Median is 48 months	⅛ live 7–14 years
Ph^1-positive with some Ph^1-negative hematopoietic cell subpopulations	—	—	—	—	—	—	—	Median is 31 months	—
Ph^1-negative (100% metaphases)	Predominantly males	100,000/mm^3	Severe thrombocytopenia	Pronounced anemia	Less marked splenomegaly	—	Poor	Median is 10–20 months	—

Source: Sandberg (1980) [137].

finding in African patients with CML, namely edema of both lower extremities as the only presenting symptom, was reported by Haddock from Ghana [139].

While CML is rare in children, both the Ph^1-positive and Ph^1-negative forms of disease have been reported [140]. Ph^1-positive disease is clinically similar to that of adults and is associated with a good prognosis. Ph^1-negative CML occurs primarily in very young children, is characterized by moderate leukocytosis, thrombocytopenia, moderate myeloid and erythroid hyperplasia, and has a poor prognosis [141].

Conditions with Chromosomal Abnormalities Predisposing to Leukemia

In addition to the described chromosomal abnormalities noted in all cell types of the leukemias, various leukemias have also been associated with Down's syndrome as well as other conditions characterized by such abnormalities. The relationship between leukemia and Down's syndrome has been shown by: (1) studies investigating the frequency of Down's syndrome among cases of leukemia compared with controls in which the excess among leukemics ranged from 6- to 100-fold [108, 142–147]; (2) studies examining the frequency of leukemia among persons with Down's syndrome compared with controls in which the excess ranged from 8 to 18 times greater than expected [148–153]; and (3) family studies examining the occurrence of leukemia among family members of persons with Down's and Down's syndrome among family members of children with leukemia. Many of the first two types of studies carried out prior to 1969 were reviewed in detail by Kessler and Lilienfeld [154].

Many of the researchers investigating the relationship between Down's syndrome and leukemia obtained the leukemia cases from death certificates of children ranging in ages from infancy to age 16 [143, 146, 147]. Three studies used cases from hospital discharge listings or cancer registries, again examining leukemics ranging from 0 to 16 years in age [108, 142, 144]. Leukemia cell type was not specified in most of these studies. The investigation by Wald et al. studied leukemia deaths of all ages and noted an excess frequency of Down's syndrome in leukemics of all ages compared with that expected [145]. Fraumeni et al. also noted an association between Down's and children with ALL and AML [155].

A number of methodologic issues must be considered when attempting to interpret these studies. Both diagnoses must have been firmly established: Down's syndrome by clinical, neurologic, mental status, and laboratory (chromosome) examination and leukemia by repeated clinical and laboratory examination (blood counts, bone marrow) in order to rule out other causes of leukopenia or leukemoid reaction or abnormal bone marrow examination. An appropriate control group must have been selected with ascertainment of Down's in this group as complete as in the case group. Most of the studies failed to use homogeneous leukemia cell-type case groups to calculate frequency of Down's or failed to determine frequency of cell-type specific leukemia in persons with Down's. While one must be aware of possible methodologic pitfalls in such studies, the several different kinds of confirmatory studies carried out in different populations (with different methods of ascertaining cases) demonstrating a strong association between Down's and leukemia, particularly ALL and AML in children, provide proof for the relationship between the two conditions.

Studies that determined the frequency of leukemia among mongoloids ascertained cases in a number of ways: hospital discharge listings [149, 151, 155], institutionalized cases in mental hospitals [150, 152], and cases reported to county mental health departments [150]. Two of the studies used no comparison group [149, 155]. Others used the expected number of leukemia deaths from national mortality rates [150, 152, 153]. Without a comparison group, it is impossible to determine whether the frequency of leukemia is greater than expected. The use of national mortality rates as a comparison could result in an underestimate of the actual mortality rate since the methods used are unlikely to provide as complete ascertainment as the methods used for obtaining the case groups. The death certificate diagnoses of the control group may not be verified with medical records, whereas the cases are selected from the discharge listings that have been validated by medical chart review.

Several types of family studies have examined the occurrence of leukemia and Down's syndrome in the same family. Some of these families have been described in detail in case reports [107, 113, 156, 157]. Another type of family study is exemplified by that carried out by Holland et al., who compared the leukemia mortality experience in parents of 1,620 children with Down's syndrome with that expected from national death rates in England and Wales. They found 3 deaths from leukemia where 1.2 were expected [150]. Several investigators have examined the frequency of Down's syndrome among sibs of leukemic children. Miller found 5 cases of Down's among 1,000 sibs compared with 1.4 expected [108].

Other chromosome disorders have been reported to be associated with increased risk of leukemia including Klinefelter's syndrome and other chromosomal disorders affecting the sex chromosome [158–161].

A number of hereditary diseases are characterized by an increased risk of leukemia. These include Fanconi's anemia, a familial bone marrow hypoplasia which is manifested in the first decade of life and is associated with various malformations, and Bloom's syndrome, an autosomal recessive disorder characterized by photosensitive telangiectasia and the same type of unstable chromosome anomalies found in Fanconi's anemia. Case reports have suggested an increased risk of leukemia in persons with these disorders [162, 163]. However, systematic studies of leukemia incidence in groups of persons with these conditions compared with controls have not been carried out, in part, no doubt, because of the rarity of these conditions.

Summary

There is a growing body of evidence linking leukemia and abnormal chromosomal conditions. Leukemics of all cell types have been found to have a variety of abnormal chromosomes. The sequence of the appearance of the abnormal chromosomes and the development of the leukemia is unknown, although family studies and case reports of some individuals who underwent chromosome analysis prior to the development of leukemia indicate that the abnormality was not inherited or congenital. Progression, change or disappearance of the abnormal chromosomes has been studied in several leukemia cases.

The significance of the chromosome abnormalities in leukemics and their relationship with other etiologic factors such as radiation, chemical exposures, and viral disorders is poorly understood. Epidemiologic case control studies examining associations between these postulated risk factors and leukemia as well as between the risk factors and the leukemic subgroups with particular chromosome abnormalities may help explain the mechanism by which a given risk factor acts to cause the leukemia. Perhaps such studies may also be able to show that the chromosome abnormality may not be causally related but may be an outcome along with the leukemia. A study of a very large population of benzene-exposed workers with serial chromosome studies might help elucidate the relationship between the development of chromosome abnormalities and the development of leukemia. Clearly this is not feasible at the present time.

The development of leukemia in persons with congenital (Down's syndrome, Klinefelter's syndrome) or hereditary (Blooms' syndrome, Fanconi's anemia) disorders characterized by inherited or early-in-life induced chromosomal abnormalities is more clear-cut. In these situations, the chromosomal abnormalities do seem to place the individual at higher risk for development of leukemia. Down's syndrome is the only one of these disorders that has been systematically studied with respect to determining the frequency of leukemia in groups of cases. It is not known what additional risk factors might cause an individual with Down's syndrome, etc., to develop leukemia. The pathophysiologic mechanism by which the condition with the abnormal chromosome(s) interacts with the risk factor(s) is also unknown. Epidemiologic studies of the role of other genetic and environmental factors in leukemogenesis in persons with these disorders would provide valuable insight into the mechanism(s) whereby environmental and other genetic factors interact with existing chromosomal disorders to produce leukemia. Alternatively genetic-epidemiologic studies of such patient populations and their families may reveal an underlying disorder leading to development of both a chromosomal abnormality and, for a small subgroup of these, leukemia as a result of additional factors or the interaction of factors.

HLA AND LEUKEMIA

The HLA or human leukocyte antigen system, the major histocompatibility complex (MHC) in man, consists of a large group of antigenic specificities, each of which may be present on the white blood cells. In 1958 and 1959, Payne and Rolfs [164] and van Rood, Eernisse, and van Leeuwen [165] described white cell agglutinins which were produced in response to multiple blood transfusions and to fetomaternal immunization. These agglutinins defined the antigenic specificities which comprise the genetic system now known as HLA [166–168]. Family studies and somatic/hybridization experiments have shown that the HLA region is on the 6th chromosome [169, 170]. Genes for some components of the complement system are also present on chromosome 6, close to the HLA region. The theory that this region appears to have a significant role in immunity is reinforced by the proximity of the genes for the complement region.

The internationally recognized components of the HLA region are genetic loci designated HLA-A, -B, -C and -D [171]. Alleles (which are individual genes that may occupy the same position or locus on a specific chromosome) of the A, B, and C loci have been distinguished by the reactions of carefully selected antisera, usually obtained from multiparous women, multitransfused individuals, or recipients of organ transplants. Alleles of HLA-D are identified by a special form of mixed lymphocyte culture (MLC) reaction which uses lymphocytes from individuals who are homozygous at the HLA-D locus as stimulating cells in the MLC reaction. Each allele is designated by a letter. As an example, HLA-A7 corresponds to the "7" allele at the A locus. HLA-A, -B, -C code for glycoproteins which are noncovalently bound to beta-2 microglobulin molecules; the HLA-C glycoprotein is less well characterized than the HLA-A and -B glycoproteins.

Over the past several years, the genetic loci have become increasingly well defined, with new specificities or splits of broader specificities continuously being discovered. Studies carried out in various populations have found the frequencies of specific HLA types to differ by ethnic group. From family studies, the co-dominant inheritance of alleles of the A, B, C, and D loci as a unit, a haplotype, has been shown. Mapping and ordering of the loci has shown that the alleles sort in a nonrandom fashion. Population studies have allowed investigators to determine gene frequencies. While most combinations of alleles occur as expected, occasionally certain alleles occur together more frequently than would be expected by chance, which is referred to as linkage-disequilibrium [166–168]. The phenotypes produced by the A, B, and C loci are serologically defined (SD) antigens.

Many studies in the mouse have shown a relationship between its transplantation antigen system region called H-2 and susceptibility to mouse leukemia induced by viruses [172–176]. A genetic locus independent of H-2, the hr (hairless) locus in the highly inbred strain HRS/J, is also closely associated with susceptibility to leukemia [177].

These genetic associations in the mouse have stimulated investigators to examine possible associations between HLA antigens and leukemia in humans. Several differences between leukemia in mouse and man must be noted before drawing analogies between the role of major histocompatibility systems in the development of leukemia in the two species. Viruses have not yet been shown to play a definite role in leukemogenesis in man as they have in the mouse (see Chapter 6). In man, genetic diversity exists for many loci whereas the mice strains that have been studied are generally homozygous at all loci in the HLA region. The patients studied have exhibited great clinical heterogeneity as to cell type of leukemia, subgroups within a given cell type, cell surface markers of the leukemic cells, factors related to clinical expression of the leukemia and survival.

Several mechanisms by which the HLA complex may relate to leukemia have been postulated. One or more genes acting together in the major histocompatibility complex (MHC) may control carcinogenesis through interaction with chemical carcinogens, viruses, or radiation. The MHC may have an immunoregulatory function, i.e., one or more genes may control a specific immune response or a nonspecific immune response or immune suppression or lymphocyte killing of tumor cells. The MHC may exert genetic control of carcinogenesis through effects on cyclic AMP, other biochemical reactions, or hormone levels. Several MHC genes may regulate a single immune response.

Several possible mechanisms may explain associations between HLA antigens and disease. The association may be causal; a gene determining a specific HLA antigen may directly cause a malignancy. As noted above, a particular gene interacting with other genes or gene products may cause a malignancy through an additive or a modifying effect or the particular gene may derepress another one or more malignancy-producing genes. A gene may interact with an environmental factor and thereby indirectly cause the malignancy. It is also possible that a gene may be linked with other genes that actually cause the malignancy. Chance alone may also be responsible for the apparent association, either because of multiple comparisons or small sample sizes. Genes at the HLA loci have not been shown to be absolutely necessary or sufficient for the development or manifestation of any disease. However, 90% of all Caucasian persons with ankylosing spondylitis have the B27 antigen and 10% do not [178–180]. It is possible that B27 is not absolutely necessary or that criteria for case definition of ankylosing spondylitis are not 100% sensitive, i.e., not all true cases are identified.

Various additional methodologic issues must be considered when the literature of HLA associations with leukemias is reviewed. Sample size of case and control groups, patient and control selection must be questioned. Are the cases in the study incident cases and representative of all disease-specific cases or do they tend to include only long-term survivors, all ages or a particular age group, previously treated or untreated individuals? Technical problems in defining HLA types can arise in some situations if lymphocytes are obtained from cases during relapse. Because sample sizes of the leukemia case groups studied are small and the leukemias are rare, some investigators have attempted to combine the results of several studies. However, the data in such combined analyses are usually heterogeneous and estimates of relative risks determined from such analyses must be interpreted cautiously.

In 1967, Amiel noted an increased frequency of antigen 4c in patients with Hodgkin's disease [181]. Other authors also noted an increased proportion of A5 and W5 (included in the antigen 4c identified by Amiel) in Hodgkin's cases compared with controls [182, 183]. More recently, the 4c antigen has been subdivided into the cross-reacting specificities B5, BW15, BW18, and BW35, and a number of investigators have noted an association between one or more of these specificities and Hodgkin's disease [184, 185]. Other investigators found an increase in A1 and B8 [186, 187]. Some authors believe that one or more HLA antigens might be associated with long-term survival [188, 189] but other evidence does not support this [190].

ALL Harris et al. [191] and Svejgaard [185] reviewed the studies of the frequency of HLA types in ALL. The numbers of cases per study ranged from 10 to 126 with the average being 50. Virtually all of the studies compared cases with controls and many of the studies examined family members. Some investigators found increased risks associated with A2 [192–194], but others did not [195–200]. A decreased risk was found for A9, that is, a smaller proportion of cases had A9 than did controls [193, 199, 201]. For those patients who were A9 positive, a longer survival time was noted [199, 202]. Klouda et al. typed incident or new onset cases of ALL and noted fewer than expected individuals with A9 at presentation which supports the theory that the presence of A9 may be associated with resistance to ALL.

Warren et al. found an increased frequency of the group 5a antigen in patients with ALL compared with controls [203, 204]. The leukocyte group 5 antigenic system, consisting of two alleles 5a and 5b, segregates independently of the HLA system and these antigens are expressed on several cells including lymphocytes. Family studies of family members and ALL index cases showed increased proportions of ALL patients with 5a compared with their siblings. The studies of Warren et al. suggest that 5a was more frequent in those patients who failed to achieve a sustained remission with chemotherapy. However, these studies combined incident and prevalent cases of ALL and therefore inferences must be made cautiously.

MacSween et al. reported a restricted genetic heterogeneity in parents of patients with ALL; 7/26 parents shared a common haplotype compared with 1/27 control parents [205]. Werner-Favre and Jeannet also reported an increased frequency of parents of children with acute leukemia sharing at least two common antigens compared with controls but the cases were not further characterized in terms of cell type of acute leukemia [206].

CLL Pollack and DuBois [207] found increased frequencies of A9, A28, and Aw29 in 34 CLL patients compared with healthy controls matched for age, sex, race, and ethnic origin. The A9 association was also reported by Jeannet and Magnin who studied 15 patients with CLL, matched with healthy individuals of the same ethnic origin and living in the same region [195]. Other studies of HLA antigen frequencies in patients with CLL have shown no notable differences between cases and controls [208]. These results must be interpreted cautiously because all the studies involved small numbers of cases from single institutions, thus raising the possibility of selection biases (i.e., the possibility that selected cases may not represent the full spectrum of disease).

Because a possible association of autoimmune disorders with CLL has been postulated, the HLA antigen association with several autoimmune diseases may provide etiologic clues. B8 has been found to be increased in chronic active hepatitis, adult celiac disease (hypersensitivity to gluten), childhood asthma, dermatitis herpetiformis, diabetes mellitus, systemic lupus erythematosus, Grave's disease, idiopathic Addison's disease, and myasthenia gravis [193, 209]. The strong association between ankylosing spondylitis, Reiter's disease, and HLA-B27 has already been described [210]. The association of HLA antigens with several autoimmune disorders, anecdotally reported to be associated with CLL, suggests that it might be fruitful to examine the relationship of HLA to CLL in the context of a well-designed, case-control epidemiologic study [211].

ANLL A number of studies have examined associations between HLA and AML as well as other types of ANLL. No significant associations were found by a number of investigators [200, 212–217]. But Heise et al. analyzed data from their series of 79 AML patients and noted an increased frequency of B17 and B27 in AML patients over 40 years of age [218]. These investigators then combined their data with that of Jeannet and Magnin [213], Hester et al. [200], and Albert et al. [217] and found that B17 was significantly but weakly associated with AML (RR = 1.48). Heise et al. also found that induction of remission occurred less frequently in patients with B17 compared to patients lacking this antigen. No associations between survival and particular HLA types were found.

Oliver et al. found an increased frequency of B12 in their series of 150 patients with AML compared with 234 normal controls [219]. In this series, patients with the antigen pairs A1 and B8 or A2 and B12 survived longer than those without either of these pairs; patients with three or more of these four antigens had longer survival than those with two or fewer.

Harris et al. found a decreased frequency of B12 at the time of presentation in cases with AML compared with controls. However, AML cases with B12 had longer survivals than those without B12 [196]. These authors found no associations between survival and the pairs A1 and B8 or A2 and B12 although persons with either of these two pairs or the pair A3 and B15 had a greater probability of going into remission after receiving induction chemotherapy.

CML While Hester et al. found an increased frequency of three antigens (B12, B7, BW35) in 48 cases with CML compared with controls, the differences were not statistically significant. No increases in any haplotype were noted. No associations were found between any antigen and survival [220].

OTHER GENETIC MARKERS

In addition to the numerous studies examining possible associations between chromosomes, HLA, and the leukemias, investigators have examined a number of other genetic markers including dermatoglyphics (ridge patterns of the skin of the fingers, palms, toes and soles), blood groups, serum proteins, haptoglobins, and enzyme alleles by comparing leukemics to controls.

Dermatoglyphic abnormalities are found in certain genetic disorders, particularly Down's syndrome, that are associated with an increased incidence of leukemia [221]. In general, dermatoglyphics are believed to be genetically determined but certain teratogens, including rubella [222, 223], and certain drugs taken in pregnancy [224], may also result in dermatoglyphic anomalies. Table 5-2 summarizes several of the studies that have attempted to examine this association. The results show great variation

Table 5-2 Dermatoglyphics in leukemia: Summary of studies

Investigator and year	Types of leukemia studied	Number of cases	Control group	Results
Aleksandrowicz et al., 1966 [225]	"Different types" of leukemia; number in subgroups by type not specified	175	600 persons, both sexes, drawn from the Cracow, Poland, region	In male leukemics only, a significant excess of radial loops; in female leukemics only, a significant excess of B whorls
Menser and Purvis-Smith, 1969 [226]	Children with leukemia, documented by bone marrow exam	25	100 children, both sexes, who were sibs of children seen as outpatients in the same hospital	Increased arches, Simian lines and Sydney lines in leukemics; decreased ulnar loops
Rosner, 1969 [227]	Adults and children with various leukemia cell types	188 total 67 AML 31 CML 60 CLL 30 ALL	271 controls (methods of identification and selection not given)	Increased radial loops in men with AML; increased whorls and decreased ulnar loops in both sexes with CML; increased arches in women with CLL
Verbov, 1969 [228]	Patients with leukemia—breakdown by cell type not given	100	None mentioned	Increased radial loops and decreased arches in men with CLL; increased arches in women with CLL
Wertelecki et al., 1969 [229]	ALL	54	Equal number of healthy, unrelated Caucasian males and females of similar age who served as controls in another study	No difference between cases and controls
Nora et al., 1969 [230]	Children with acute leukemia	56	Two groups: (1) First-degree relatives; (2) Unrelated individuals	No difference between cases and two control groups
Zahalkova and Belusa, 1970 [231]	Children with acute leukemia	20	40 men and women from same geographic area	Statistically significant increased incidence of Sydney lines

and are difficult to interpret. Several of the studies grouped the leukemias together and did not report the findings by cell type. Some used small sample sizes and reported results from 30 or fewer leukemics. One study [225] did not use controls. When controls were used, the method of control selection varied from study to study. Numerous possible lines, angles, patterns on palms and fingers can be measured and compared. It is likely that by chance alone some would be found to be associated with leukemia. Combining groups of leukemics by sex or race may be a methodologic problem because dermatoglyphic patterns are known to vary with sex and race. Because the findings were inconsistent for palm and finger lines, whorls, etc., evidence does not support a conclusion that dermatoglyphic characteristics may be significant as markers for leukemia [225–231].

Investigations of blood groups, serum proteins, haptoglobins, and enzyme alleles (all of which are genetically determined characteristics) have, in general, not shown any major differences between leukemics and controls. These studies have been validly criticized. Most of them have not analyzed separately the different cell types. In certain instances, leukemias were grouped with lymphomas. In some studies, control groups were selected from populations that differed genetically from populations to which the affected subjects belonged. Published papers are weighted heavily toward positive findings. In general, there was often failure to publish negative results. Finally, sample sizes are often quite small.

Both acute and chronic leukemics (types not further specified) are slightly more likely to have blood group A than are controls (acute leukemia RR = 1.12; chronic leukemia RR = 1.08) [232, 233].

Lather et al. [234], Pencock [235], and Wendt et al. [236] found that the leukemics (not further analyzed by cell type) had an excess of individuals homozygous for haptoglobin type 1-1 and a deficiency of persons homozygous for type 2-2 compared with controls. However, Galatius-Jensen found no differences in haptoglobin types between leukemics and controls [237].

Gurda noted that persons with AML are homozygous for type 2-2 of the polygraphic blood marker GC much more commonly than controls. This difference was not found for CML or CLL [238].

The significance of differences in the distribution of most genetic markers between leukemics and controls is not clear. Such differences (most of which are small or equivocal) must be interpreted cautiously because of the problems described above.

Addendum

Recently techniques have become available for cytogenetic analysis of malignant B lymphocytes [239]. Prior to this, as indicated earlier in the chapter, cytogenetic studies of phytohemagglutinin-induced metaphases of peripheral lymphocytes in CLL patients were reported to have normal karyotypes in almost all cases [240, 241]. However, a number of investigators have now reported cytogenetic abnormalities in a substantial proportion of patients with CLL who have been studied using techniques involving B-cell mitogens and combined T and B mitogens [74, 84, 239, 242–246]. A number of abnormal karyotypes have been found including trisomy 12 (the most common), trisomy 16, trisomy 18, and others. Han et al. have also found that abnormal karyotypes were significantly associated with advanced clinical stage and shortened survival [239].

REFERENCES

1. F.W. Gunz, "The epidemiology and genetics of the chronic leukemias," *Clin. Haematol.* **6:**3–20 (1977).
2. C.W. Heath, Jr., "The epidemiology of leukemia," in *Cancer Epidemiology and Prevention: Current Concepts,* pp. 318–350. (D. Schottenfeld, ed). Springfield, Illinois: Charles C. Thomas, 1975.
3. S. Kurita and Y. Kamei, "Genetics of familial leukemia," *Jpn. J. Hum. Genet.* **14:**163–179 (1969).
4. J. Cervenka, R.S. Anderson, M.E. Nesbit, and W. Krivit, "Familial leukemia and inherited chromosomal aberration" *Int. J. Cancer* **19:**783–788 (1977).
5. H.W.L. Hann, W.T. London. A.I. Sutnick, B.S. Blumberg. E. Lustbader, H.M. Carim, A.E. Evans, H.E.M. Kay, and I.C.M. Maclennan. "Studies of parents of children with acute leukemia," *J. Natl. Cancer Inst.* **54:**1299–1305 (1975).
6. W.A. Blattner, W. Strober, A.V. Muchmore, R.M. Blaese, S. Broder, and J.F. Fraumeni, Jr., "Familial chronic lymphocytic leukemia," *Ann. Intern Med.* **84:**554–557 (1976).
7. F.W. Gunz, J.P. Gunz, P.C. Vincent, M. Bergin, F.L. Johnson, H. Bashir, and R.L. Kirk, "Thirteen cases of leukemia in a family," *J. Natl. Cancer Inst.* **60:**1243–1250 (1978).
8. R. Goudsmit, A.M. v. Leeuwen, and J. James, "Dohle bodies and acute myeloblastic leukaemia in one family: a new familial disorder?" *Br. J. Haematol.* **20:**557–561 (1971).
9. F. Gunz and A.G. Baikie, "Leukemia in the past," in *Leukemia,* 3rd ed., pp. 3–12. New York: Grune & Stratton, 1974.
10. A. Decastello, "Beitrag zur Kenntnis der familiaren Leukamie," *Med. Klin.* **35:**1255–1257 (1939).
11. J.H. Hornbaker, "Chronic leukemia in three sisters," *Am. J. Med. Sci.* **203:**322–325 (1942).

12. A. Videbaek, "Familial leukemia. A preliminary report," *Acta Med. Scand.* **127**:26–52 (1947).
13. R.C. Anderson, "Familial leukemia. A report of leukemia in five siblings with a brief review of the genetic aspects of the disease," *Am. J. Dis. Child.* **81**:313–322 (1951).
14. E.B. Reilly, S.I. Rapaport, N.W. Karr, H. Mills, and G.E. Carpenter, "Familial chronic lymphatic leukemia," *Arch. Intern. Med.* **90**:87–89 (1952).
15. F.W. Gunz and W. Dameshek, "Chronic lymphocytic leukemia in a family including twin brothers and a son," *J. Am. Med. Assoc.* **164**:1323–1325 (1956).
16. M.S.E. Johnson and C.H. Peters, "Lymphoma in four siblings," *J. Am. Med. Assoc.* **163**:20–25 (1957).
17. A.G. Steinberg, "The genetics of acute leukemia in children," *Cancer* **13**:985–999 (1960).
18. W.A.B. Campbell, A.L. Macafee and W.G. Wade, "Familial neonatal leukaemia," *Arch. Dis. Child.* **37**:93–98 (1962).
19. R.D. Gordon, "Hereditary factors in human leukaemia: report of four cases of leukaemia in a family," *Aust. Ann. Med.* **12**:202–207 (1963).
20. D.L. Randall, C.W. Reiquam, J.H. Githens and A. Robinson, "Familial myeloproliferative disease. A new syndrome closely simulating myelogenous leukemia in childhood," *Am. J. Dis. Child.* **110**:479–500 (1965).
21. C.W. Heath, Jr. and W.C. Moloney, "Familial leukemia. Five cases of acute leukemia in three generations," *N. Engl. J. Med.* **272**:882–887 (1965).
22. F.W. Gunz, P.H. Fitzgerald, P.E. Crossen, I.S. MacKenzie, C.P. Powles and G.R. Jensen, "Multiple cases of leukemia in a sibship," *Blood* **27**:482–489 (1966).
23. P. McPhedran, C.W. Heath, Jr. and J. Lee, "Patterns of familial leukemia, Ten cases of leukemia in two interrelated families," *Cancer* **24**:403–407 (1969).
24. F.W. Gunz and A.M.O. Veale, "Leukemia in close relatives—accident or predisposition?" *J. Natl. Cancer Inst.* **42**:517–524 (1969).
25. J.F. Fraumeni, Jr.. C.L. Vogel and V.T. DeVita, "Familial chronic lymphocytic leukemia," *Ann. Intern Med.* **71**:279–284 (1969).
26. S.W. Ferguson and T.N. Lynn, "Familial leukemia: a report of four cases of acute leukemia in four consecutive generations," *South. Med. J.* **63**:1337–1340 (1970).
27. A.L. Snyder, F.P. Li, E.S. Henderson and G.J. Todaro, "Possible inherited leukaemogenic factors in familial acute myelogenous leukemia," *Lancet* **1**:586–589 (1970).
28. A.I. Potolsky, C.W. Heath, Jr., C.E. Buckley, III and D.T. Row-

28. A.I. Potolsky, C.W. Heath, Jr., C.E. Buckley, III and D.T. Rowlands, Jr., "Lymphoreticular malignancies and immunologic abnormalities in a sibship," *Am. J. Med.* **50**:42–48 (1971).
29. M.L. Ghosh, "Familial leukaemia," *Acta Haematol. (Basel)* **48**:98–103 (1972).
30. T.W. Pendergrass, D.L. Mann, R.G. Stoller, R.H. Halterman and J.F. Fraumeni, Jr., "Acute myelocytic leukaemia and leukaemia-associated antigens in sisters," *Lancet* **2**:429–430 (1975).
31. P. McPhedran and C.W. Heath, Jr., "Multiple cases of leukemia associated with one house," *J. Am. Med. Assoc.* **209**:2021–2025 (1969).
32. A. Videbaek, "Heredity in human leukemia and its relation to cancer: a genetic and clinical study of 209 probands," Thesis, Copenhagen: Eijnar Munksgaard, 1947.
33. P.L. Amiotti, "Sulla incidenza dei tumori nei familiari di bambini leucemici," *Minerva Pediatr.* **5**:449–450 (1953).
34. G. Morganti and A. Cresseri, "Nouvelles recherches génétiques sur les leucémies," *Sang* **25**:448–453 (1954).
35. C. Kaliampetsos, "Kommen Blut Krankheiten und Karzinome unter den Verwandten von leukämiekranken gehäuft vor?" *Dtsch. Med. Wschr.* **79**:1773–1785 (1954).
36. F.W. Gunz. "Leukemia in New Zealand and Australia," *Pathol. Microbiol.* **27**:697–704 (1964).
37. F.W. Gunz, J.P. Gunz, A.M.O. Veale, C.J. Chapman and I.B. Houston, "Familial leukaemia: a study of 909 families," *Scand. J. Haematol.* **15**:117–131 (1975).
38. B. MacMahon and M. Levy, "Prenatal origin of childhood leukemia: evidence from twins," *N. Engl. J. Med.* **270**:1082–1085 (1964).
39. H. Jelke, "Akute lymphatische Leukämie bei eineiigen Zwillingen," *Acta Paediatr.* **27**:87–136 (1939).
40. L. Riel, "Akute lymphatische leukämie bei (eineiigen) Zwillingen," *Kinderärztl. Prax.* **16**:148–151 (1948).
41. R.C. Anderson and H.W. Hermann, "Leukemia in twin children," *J. Am. Med. Assoc.* **158**:652–654 (1955).
42. R. Wegelius and E.J. Paaso, "Identical twins simultaneously dying of leukemia," *Ann. Paediatr. Fenn.* **2**:137–141 (1956).
43. A.S. Tikhomirov and B.Z. Sirotin, "Case of leukaemia in two uniovular twins," *Probl. Hematol. Blood Trans (U.S.S.R.)* **4**:75–79 (1959).
44. H. Ioachim, "Acute leukemia in uniovular twins: review of genetic aspects of human leukemia," *Cancer* **15**:539–545 (1962).
45. H.A. Pearson, F.W. Grello and T.E. Cone, Jr., "Leukemia in identical twins," *N. Engl. J. Med.* **268**:1151–1156 (1963).

46. W.W. Zuelzer and D.E. Cox, "Genetic aspects of leukemia," *Semin. Hematol.* **6:**4–25 (1969).
47. W.M. Court-Brown and R. Doll, "Leukemia in childhood and young adult life. Trends in mortality in relation to aetiology," *Br. Med. J.* **1:**981–988 (1961).
48. R.W. Miller, "Death from childhood cancer in sibs," *N. Engl. J. Med.* **279:**122–126 (1968).
49. R.W. Miller, "Deaths from childhood leukemia and solid tumors among twins and other sibs in the United States, 1960–67," *J. Natl. Cancer Inst.* **46:**203–209 (1971).
50. J.G. Feldman, S.L. Lee and B. Seligman, "Occurrence of acute leukemia in females in a genetically isolated population," *Cancer* **38:**2548–2550 (1976).
51. P.C. Nowell and D.A. Hungerford, "A minute chromosome in human chronic granulocytic leukemia," *Science* **132:**1497 (1960).
52. A.A. Sandberg, "The leukemias," in *The Chromosomes in Human Cancer and Leukemia,* pp. 183–353. New York: Elsevier, 1980.
53. T. Caspersson, L. Zech, E.J. Modest, G.E. Foley, U. Wagh and E. Simonsson, "DNA-binding fluorochromes for the study of the organization of the metaphase nucleus," *Exp. Cell Res.* **58:**141–152 (1969).
54. S.D. Lawler, L.M. Secker Walker, B.M. Summersgill, B.R. Reeves, J. Lewis, H.E.M. Kay and R.M. Hardisty, "Chromosome banding studies in acute leukemia at diagnosis," *Scand. J. Haematol.* **15:**312–320 (1975).
55. J.J. Yunis, C.D. Bloomfield and K. Ensrud, "All patients with acute nonlymphocytic leukemia may have a chromosome defect," *N. Engl. J. Med.* **305:**135–139 (1981).
56. P.C. Nowell, "Marrow chromosome studies in 'preleukemia.' Further correlation with clinical course," *Cancer* **28:**513–518 (1971).
57. P. Nowell and J. Finan, "Chromosome studies in preleukemic states. IV. Myeloproliferative versus cytopenic disorders," *Cancer* **42:**2254–2261 (1978).
58. R.V. Pierre, "Cytogenetic studies in preleukemia. Studies before and after transition to acute leukemia in 17 subjects," in *Unclassifiable Leukemias,* pp. 163–172. Berlin: Springer-Verlag, 1975.
59. R. Lisker, A.C. Gutierrez and M. Velasquez-Ferrari, "Longitudinal bone marrow chromosome studies in potential leukemic myeloid disorders," *Cancer* **31:**509–515 (1973).
60. A. Panani, A.G. Papayannis and E. Sioula, "Chromosome aberrations and prognosis in preleukaemia," *Scand. J. Haematol.* **24:**97–100 (1980).
61. M.C. Cimino, J.D. Rowley, A. Kinnealey, D. Variakojis and H.M.

Golomb, "Banding studies of chromosomal abnormalities in patients with acute lymphocytic leukemia," *Cancer Res.* **39:**227–238 (1979).

62. A.A. Sandberg. N. Takagi, T. Sofuni and L.H. Crosswhite, "Chromosomes and causation of human cancer and leukemia. III. Karyotypic aspects of acute leukemia," *Cancer* **22:**1268–1282 (1968).
63. A.A. Sandberg, T. Ishihara, Y. Kikuchi and L.H. Crosswhite, "Chromosomal differences among the acute leukemias," *Ann. N.Y. Acad. Sci.* **113:**663–716 (1964).
64. M. Oshimura, A.I. Freeman and A.A. Sandberg, "Chromosomes and causation of human cancer and leukemia. XXVI. Banding studies in acute lymphoblastic leukemia (ALL)," *Cancer* **40:**1161–1171 (1977).
65. M. Oshimura and A.A. Sandberg, "Chromosomal 6q-anomaly in acute lymphoblastic leukemia," *Lancet* **2:**1045–1046 (1976).
66. W.W. Zuelzer, S. Inoue, R.I. Thompson and M.J. Ottenbreit, "Long-term cytogenetic studies in acute leukemia of children: The nature of relapse," *Am. J. Hematol.* **1:**143–190.
67. J. Whang-Peng, T. Knutsen, J. Ziegler and B. Leventhal, "Cytogenetic studies in acute lymphocytic leukemia: special emphasis on long-term survival," *Med. Pediatr. Oncol.* **2:**333–351 (1976).
68. K. Sohn and D. Boggs, "Klinefelter's syndrome, LSD usage, and acute lymphoblastic leukemia," *Clin. Genet.* **6:**20–22 (1974).
69. A. Kessous, J. Corberand, A. Robert and P. Colombiés, "Caractéres cytogénétiques des leucémies aiguës de l'enfant," *Lyon Med.* **233:**253–260 (1975).
70. E. Hinkes, B.F. Crandall, F. Weber and C.G. Craddock, "Acute leukemia with C-G chromosome translocation," *Blood* **41:**259–263 (1973).
71. F.G.J. Hayhoe, L.F. Sinks and R.J. Flemans, "Studies on the transformation *in vitro* of lymphocytes from chronic lymphocytic leukemia," in *The Lymphocyte in Immunology and Haemopoiesis* (Symposium, Bristol, 1966), pp. 66–72. London: Edward Arnold, 1966.
72. D. Quaglino and D.C. Cowling, "Cytochemical studies on cells from chronic lymphocytic leukaemia and lymphosarcoma cultured with phytohaemagglutinin," *Br. J. Haematol.* **10:**358–364 (1964).
73. P.H. Fitzgerald and A. Adams, "Chromosome studies in chronic lymphocytic leukemia and lymphosarcoma," *J. Natl. Cancer Inst.* **34:**827–839 (1965).
74. G. Gahrton, K-H. Robert, K. Fiberg and L. Zech, "Extra chromosome 12 in chronic lymphocytic leukemia," *Lancet* (Letter to the Editor) **1:**146–147 (1980).
75. I. Macavei and S. Halmos, "High T-cell counts in chronic lymphatic leukemia," *Lancet* **1:**220–221 (1975).

76. E. König, G. Cohnen, G. Brittinger and S.D. Douglas, "Response to phytohaemagglutinin and pokeweed mitogen in chronic lymphocytic leukemia," *Lancet* (Letter to the Editor) **1**:794–795 (1972).
77. H.J. Woodliff and G. Cohen, "Cytogenetic studies in chronic lymphocytic leukemia. I. A study of 40 patients," *Med. J. Aust.* **1**:970–974 (1972).
78. F.W. Gunz, P.H. Fitzgerald and A. Adams, "An abnormal chromosome in chronic lymphocytic leukaemia," *Br. Med. J.* **2**:1097–1099 (1962).
79. P.H. Fitzgerald and F.W. Gunz, "Chromosomal abnormality and chronic lymphocytic leukemia," *Lancet* (Letter to the Editor) **2**:150 (1964).
80. P.H. Fitzgerald, P.E. Crossen, P.E. Adams, C.V. Sharman and F.W. Gunz, "Chromosome studies in familial leukaemia," *J. Med. Genet.* **3**:96–100 (1966).
81. W.M. Court-Brown, "Chromosomal abnormality and chronic lymphatic leukemia," *Lancet* (Letter to the Editor) **1**:986 (1964).
82. C. Abbot, "The Christchurch chromosome," *Lancet* **1**:1155–1156 (1966).
83. J.A. Sonnier, G.R. Buchanan, P.N. Howard-Peebles, J. Rutledge and R. Graham Smith, "Chromosomal translocation involving the immunoglobulin kappa-chain and heavy-chain loci in a child with chronic lymphocytic leukemia," *N. Engl. J. Med.* **309**:590–594 (1983).
84. G. Gahrton and K-H. Robert, "Chromosomal aberrations in chronic B-cell lymphocytic leukemia," *Cancer Genet. Cytogenet.* **6**:171–181 (1982).
85. G.M. Lenoir, J.L. Preudhomme, A. Bernheim and R. Berger, "Correlation between immunoglobulin light chain expression and variant translocation in Burkitt's lymphoma and leukemia," *Nature* **294**:313–318 (1981).
86. M. Sakurai and A.A. Sandberg, "Chromosomes and causation of human cancer and leukemia. XI. Correlation of karyotypes with clinical features of acute myeloblastic leukemia," *Cancer* **37**:285–299 (1976).
87. J.D. Rowley and D. Potter, "Chromosomal banding patterns in acute nonlymphocytic leukemia," *Blood* **47**:705–721 (1976).
88. J.D. Rowley, "Missing sex chromosomes and translocation in acute leukemia," *Lancet* **2**:835–836 (1974).
89. P.H. Fitzgerald and J.W. Hamer, "Karyotype and survival in human acute leukemia," *J. Natl. Cancer Inst.* **56**:459–462 (1976).
90. H. Van den Berghe, G.H. Borgström, L. Brandt, D. Catovksy, A. de la Chapelle, H.M. Golomb, D.K. Hossfeld, S. Lawler, J. Lind-

sten, A. Louwagie, F. Mitelman, B.R. Reeves, J.D. Rowley, A.A. Sandberg, L. Teerenhovi and P. Vuopio, "Chromosomes in acute non-lymphocytic leukemia," *Br. J. Haematol.* **39**:311–316 (1978).

91. F. Mitelman and L. Brandt, "Chromosome banding pattern in acute myeloid leukemia," *Scand. J. Haematol.* **13**:321–330 (1974).
92. J.D. Rowley, H.M. Golomb and C. Dougherty, "15/17 translocation, a consistent chromosomal abnormality in acute promyelocytic leukemia," *Lancet* **1**:549–550 (1977).
93. P. Philip, M.K. Jensen, S.A. Killman, A. Drivsholm and N.E. Hansen, "Chromosomal banding patterns in 88 cases of acute nonlymphocytic leukemia," *Leukemia Res.* **2**:201–212 (1978).
94. H.M. Golomb, J.W. Vardiman, J.D. Rowley, J.R. Test and V. Mintz, "Correlation of clinical findings with quinacrine-banded chromosomes in 90 adults with acute nonlymphocytic leukemia: an eight-year study (1970–77)," *N. Engl. J. Med* **299**:613–619 (1978).
95. M. Sakurai and A.A. Sandberg, "Prognosis of acute myeloblastic leukemia: chromosomal correlation," *Blood* **41**:93–104 (1973).
96. P.G. Nilsson, L. Brandt and R. Mitelman, "Prognostic implications of chromosome analysis in acute non-lymphocytic leukemia," *Leukemia Res.* **1**:31–34 (1977).
97. W.F. Benedict, M. Lange, J. Greene, A. Derencsenyi and O.S. Alfi, "Correlation between prognosis and bone marrow chromosomal patterns in children with acute nonlymphocytic leukemia: similarities and differences compared to adults," *Blood* **54**:818–823 (1979).
98. D.A.G. Galton and J.V. Dacie, "Classification of the acute leukemias," *Blood Cells* **1**:17–24 (1975).
99. First Internatonal Workshop in Chromosomes in Leukemia, *Cancer Res.* **38**:867–868 (1978).
100. H. Van den Berghe, A. Louwagie, A. Broeckaert-Van Orshoven, G. David, R. Verwilgen, J.L. Michaux, A. Ferrant and G. Sokal, "Chromosome abnormalities in acute promyelocytic leukemia (APL)," *Cancer* **43**:558–562 (1979).
101. L. Teerenhovi, G.H. Borgstrom, F. Mitelman, L. Brandt, P. Vuopio, T. Timonen, A. Almqvist and A. de la Chapelle, "Uneven geographical distribution of 15:17-translocation in acute promyelocytic leukaemia," *Lancet* (Letter to the Editor) **2**:797 (1978).
102. F.W. Gunz, R.B.M. Ravich, P.C. Vincent, J.H. Stewart, P.E. Crossen and J. Mellor, "A case of acute leukemia with rapidly changing chromosome constitution," *Ann. Genet. (Paris)* **13**:79–86 (1970).
103. L. Koulischer, "Mitotic chromatid separation in a case of acute leukemia," *Eur. J. Cancer* **2**:347–352 (1966).
104. E.W. Houston, W.C. Levin and S.E. Ritzmann, "Endoreduplication in untreated early leukaemia," *Lancet* **2**:496–497 (1964).

105. J.M. Trujillo, A. Cork, M.J. Ahearn, E.L. Youness and K.B. McCredie, "Hematological and cytological characterization of the 8/21 translocation in acute leukemia," *Blood* **53:**695–706 (1979).
106. A.A. Sandberg, J. Cortner, M. Takagi, M.A. Moghadam and L.H. Crosswhite, "Differences in chromosome constitution of twins with acute leukemia," *N. Engl. J. Med.* **275:**809–812 (1966).
107. K.E. Buckton, D.G. Harnden, A.G. Baikie and G.E. Woody, "Mongolism and leukaemia in the same sibship," *Lancet* **1:**171–172 (1961).
108. R.W. Miller, "Down's syndrome (mongolism), other congenital malformations and cancers among sibs of leukemic children," *N. Engl. J. Med.* **268:**393–401 (1963).
109. M.W. Thompson, R.E. Bell and A.S. Little, "Familial 21-trisomic mongolism coexistent with leukemia," *Can. Med. Assoc. J.* **88:**893–894 (1963).
110. K. Goh, H. Lee and G. Miller, "Down's syndrome and leukemia: mechanism of additional chromosomal abnormalities," *Am. J. Ment. Defic.* **82:**542–548 (1978).
111. C.W. Heath, Jr. and W.C. Moloney, "Cytogenetic observations in a case of erythremic myelosis," *Cancer* **18:**1495–1504 (1965).
112. P.E. Conen and B. Erkman, "Combined mongolism and leukemia. Report of eight cases with chromosome studies," *Am. J. Dis. Child.* **112:**429–443 (1966).
113. A.J. Ebbin, C.W. Heath, Jr., R.E. Moldow and J. Lee, "Down's syndrome and leukemia in a family," *J. Pediatr.* **73:**917–920 (1968).
114. F. Rosner and S.L. Lee, "Down's syndrome and acute leukemia: myeloblastic or lymphoblastic? Report of forty-three cases and review of the literature," *Am. J. Med.* **53:**203–218 (1972).
115. H.E. Sutton, "Chromosomal abnormalities: numerical," in *An Introduction to Human Genetics,* pp. 59–78. New York: Holt, Rinehart and Winston, 1975.
116. J. Bauke, "Cytogenetic studies of a patient with chronic myelocytic leukemia and his non-leukemic identical twin," *Acta Genet. Med. Gemellol. (Rome)* **19:**180–182 (1970).
117. K.O. Goh and S.N. Swisher, "Identical twins and chronic myelocytic leukemia. Chromosomal studies of a patient with chronic myelocytic leukemia and his normal twin," *Arch. Intern. Med.* **115:**475–478 (1965).
118. H.J. Woodliff, L. Dougan and P. Onesti, "Cytogenetic studies in twins, one with chronic granulocytic leukaemia," *Nature* **211:**533 (1966).
119. K.O. Goh, S.N. Swisher and E.C. Herman, "Chronic myelocytic leukemia and identical twins. Additional evidence of the Philadelphia

chromosome as postzygotic abnormality," *Arch. Intern. Med.* **120:**214–219 (1967).

120. G.K. Tokuhata, C.L. Neely and D.L. Williams, "Chronic myelocytic leukemia in identical twins and a sibling," *Blood* **31:**216–225 (1968).
121. A.K. Maniatis, S. Amsel, W.J. Mitus and N. Coleman, "Chromosome pattern of bone marrow fibroblasts in patients with chronic granulocytic leukemia," *Nature* **222:**1278–1279 (1969).
122. A.G. Baikie, O.M. Garson, A.S.D. Spiers and J. Ferguson, "Cytogenetic studies in familial leukemias," *Aust. Ann. Med.* **18:**7–11 (1969).
123. S.I. Sonta and A.A. Sandberg, "Chromosomes and causation of human cancer and leukemia. XXIV. Unusual and complex Ph[1]-translocations and their clinical significance," *Blood* **50:**691–697 (1977).
124. P.J. Fialkow, S.M. Gartler and A. Yoshida, "Clonal origin of chronic myelocytic leukemia in man," *Proc. Natl. Acad. Sci. USA* **58:**1468–1471 (1967).
125. R.D. Barr and P.J. Fialkow, "Clonal origin of chronic myelocytic leukemia," *N. Engl. J. Med.* **289:**307–309 (1973).
126. P.H. Fitzgerald, A.F. Pickering and J.R. Elby, "Clonal origin of the Philadelphia chromosome and chronic myeloid leukemia: evidence from a sex chromosome mosaic," *Br. J. Haematol.* **21:**473–480 (1971).
127. J.W. Adamson and P.J. Fialkow, "Annotation. The pathogenesis of myeloproliferative syndromes," *Br. J. Haematol.* **38:**299–303 (1978).
128. P.C. Nowell, "Tumors as clonal proliferation," *Virchows Archiv. B.* [Cell Pathol.] **29:**145–150 (1978).
129. I.M. Tough, W.M. Court-Brown, A.G. Baikie, K.E. Buckton, D.G. Harden, P.A. Jacobs, M.J. King and J.A. McBride, "Cytogenetic studies in chronic myeloid leukemia and acute leukemia associated with mongolism," *Lancet* **1:**411–417 (1961).
130. D.K. Hossfeld, T. Han, R.N. Holdsworth and A.A. Sandberg, "Chromosomes and causation of human cancer and leukemia: VII. The significance of the Ph[1] in conditions other than CML," *Cancer* **27:**186–192 (1971).
131. I. Hayata, S. Kakati and A.A. Sandberg, "A new translocation related to the Philadelphia chromosome," *Lancet* **2:**1385 (1973).
132. S.D. Lawler, D.S. Lobb and E. Wiltshaw, "Philadelphia chromosome positive bone-marrow cells showing loss of the Y in males with chronic myeloid leukemia," *Br. J. Haematol.* **27:**247–252 (1974).
133. A.A. Horland, S.R. Wolman, A. Distenfeld and T. Cohen, "Another variant translocation in chronic myelogenous leukemia," *N. Engl. J. Med.* **294:**164–165 (1976).

134. E. Engle, B.J. McCee, J.M. Flexner et al., "Philadelphia chromosome (Ph^1) translocation in an apparently Ph^1 negative, minus G22, case of chronic myeloid leukemia," *N. Engl. J. Med.* **291:**154 (1974).
135. A.A. Sandberg and M. Sakurai, "The missing Y chromosome and human leukaemia," *Lancet* (Letter to the Editor) **1:**375 (1973).
136. J. Bauke, "Extramedullary manifestations of chronic myelogenous leukemia. Presence of hyperdiploid stemlines with Ph^1 disomy in lymph nodes and pleural effusions," Meeting of the Asian-Pacific Division of the International Society of Haematology, Melbourne, 1971.
137. A.A. Sandberg, "Chronic myelocytic leukemia," in *The Chromosomes in Human Cancer and Leukemia,* pp. 205–206. New York: Elsevier, 1980.
138. M.R. Leibowitz, D.P. Derman, R. Jacobson, K. Stevens and J. Katz, "Chronic myeloid leukaemia in South African blacks," *S. Afr. Med. J.* **50:**2035–2037 (1976).
139. D.R.W. Haddock, "The pattern of leukemia in Accra, Ghana," *J. Trop. Med. Hyg.* **70:**60–62 (1967).
140. J. Wahrman, R. Voss, T. Shapiro and A. Ashkenazi, "The Philadelphia chromosome in two children with chronic myeloid leukemia," *Isr. J. Med. Sci.* **3:**380–391 (1967).
141. K.L. Smith and W. Johnson, "Classification of chronic myelocytic leukemia in children," *Cancer* **34:**670–679 (1974).
142. G. Knox, "The epidemiology of childhood leukaemia in Northumberland and Durham," *Br. J. Prev. Soc. Med.* **18:**17–24 (1964).
143. A. Stewart, J. Webb and D. Hewitt, "A survey of childhood malignancies," *Br. Med. J.* **1:**1495–1508 (1958).
144. T. Iverson, "Leukemia in infancy and childhood," *Acta Paediatr. Scand. Suppl.* **167:**50–53 (1966).
145. N. Wald, W.H. Borges, C.C. Li, J.H. Turner and M.C. Harnois, "Leukaemia associated with mongolism," *Lancet* **1:**1228 (1961).
146. R. Barber and P. Spiers, "Oxford survey of childhood cancers; progress report II," *Mon. Bull. Min. Health Public Health Lab. Serv.* **23:**46–52 (1964).
147. E.A. Ager, L.M. Schuman, H.M. Wallace, A.B. Rosenfield and W.H. Gullen, "An epidemiological study of childhood leukemia," *J. Chronic Dis.* **18:**113–132 (1965).
148. D.H. Merrit and J.S. Harris, "Mongolism and acute leukemia," *A.M.A. J. Dis. Child.* **92:**41–44 (1956).
149. C.O. Carter, "A life-table for mongols with the causes of death," *J. Ment. Defic.* **2:**64–74 (1958).
150. W.W. Holland, R. Doll and C.O. Carter, "The mortality from leukaemia and other cancers among patients with Down's syndrome and among their parents," *Br. J. Cancer* **16:**177–184 (1962).

151. E. DeWolff, "Etude clinique de 134 mongoliens," *Ann. Paediatr. Suppl.* **202:**1–47 (1964).
152. E.W. Jackson, J.H. Turner, M.R. Klauber and F.D. Norris, "Down's syndrome: variation of leukemia occurrence in institutionalized populations," *J. Chronic Dis.* **21:**247–253 (1968).
153. R.W. Miller, "Neoplasia and Down's syndrome," *Ann. N.Y. Acad. Sci.* **171:**637–644 (1970).
154. I.I. Kessler and A.M. Lilienfeld, "Perspectives in the epidemiology of leukemia," *Adv. Cancer Res.* **12:**225–302 (1969).
155. J.F. Fraumeni, Jr., M.D. Manning and W.J. Mitus, "Acute childhood leukemia: epidemiologic study by cell type of 1,263 cases at the Children's Cancer Research Foundation in Boston, 1947–65," *J. Natl. Cancer Inst.* **46:**461–470 (1971).
156. P.E. Conen, B. Erkman and B. Laski, "Chromosome studies on a radiographer and her family: report of one case of leukemia and two cases of Down's syndrome," *Arch. Intern. Med.* **117:**125–132 (1966).
157. O.J. Miller, W.R. Breg, R.D. Schmickel and W. Tretter, "A family with an XXXXY male, a leukaemic male, and two 21 trisomic mongoloid females," *Lancet* **2:**78–79 (1961).
158. J.F. Fraumeni, Jr. and R.W. Miller, "Epidemiology of human leukemia: recent observations," *J. Natl. Cancer Inst.* **38:**593–605 (1967).
159. J.F. Fraumeni, Jr., "Clinical epidemiology of leukemia," *Semin. Hematol.* **6:**26–36 (1969).
160. A.G. Baikie, K.E. Buckton, W.M. Court-Brown and D.G. Harnden, "Two cases of leukaemia and a case of sex chromosome abnormality in same sibship," *Lancet* **2:**1003–1004 (1961).
161. R.W. Miller, "Relation between cancer and congenital defects: an epidemiologic evaluation," *J. Natl. Cancer Inst.* **40:**1079–1085 (1968).
162. A. Sawitsky, D. Bloom and J.L. German, "Chromosomal breakage and acute leukemia in congenital telangiectatic erythema and stunted growth," *Ann. Intern. Med.* **65:**489–495 (1966).
163. G.E. Bloom, S. Warner, P.S. Gerland and L.K. Diamond, "Chromosome abnormalities in constitutional aplastic anemia," *N. Engl. J. Med.* **274:**8–14 (1966).
164. R. Payne and M.R. Rolfs, "Fetomaternal leukocyte incompatibility," *J. Clin. Invest.* **37:**1756–1763 (1958).
165. J.J. van Rood, J.G. Eernisse and A. van Leeuwen, "Leucocyte antibodies in sera from pregnant women," *Nature* **181:**1735–1736 (1958).
166. W.B. Bias, "Genetic polymorphisms and human disease," in *Bio-*

cultural Aspects of Disease, pp. 95–131. (H. Rothschild, ed.) New York: Academic Press, 1981.
167. V.A. McKusick, *Mendelian Inheritance in Man,* 5th ed. Baltimore: Johns Hopkins University Press, 1978.
168. F.H. Bach and J.J. van Rood, "The major histocompatibility complex—genetics and biology," *N. Engl. J. Med.* **295:**806–813, 872–878, 927–936 (1976).
169. A. Jongsma, H. van Someren, A. Westerveld, A. Hagemeijer and P. Pearson, "Localization of genes on human chromosomes by studies of human–Chinese hamster somatic cell hybrids," *Humangenetik* **20:**195–202 (1973).
170. L.U. Lamm, U. Friedrich, G.B. Petersen, J. Jorgensen, J. Nielsen, A.J. Therkelsen and F. Kissmeyer-Nielsen, "Assignment of the major histocompatibility complex to chromosome 6 in a family with a pericentric inversion," *Hum. Hered.* **24:**273–284 (1974).
171. WHO-IUIS Terminology Committee Nomenclature for Factors of the HLA System, *Transplant. Proc.* **8:**109–114 (1976).
172. P.A. Gorer, "The genetic and antigenic basis of tumor transplantation," *J. Pathol.* **44:**691–697 (1937).
173. G.D. Snell, "Methods for the study of histocompatibility genes," *J. Genet.* **49:**87–108 (1948).
174. F. Lilly and T. Pincus, "Genetic control of murine viral leukemogenesis," in *Advances in Cancer Research,* Vol. 17 pp. 231–277. (G. Klein, S. Weinhouse, and A. Haddow, eds.) New York: Academic Press, 1973.
175. H.S. Kaplan, "On the natural history of the murine leukemias: presidential address," *Cancer Res.* **27:**1325–1340 (1967).
176. D.C. Shreffler, C.S. David, H.C. Passmore and J. Klein, "Genetic organization and evolution of the mouse H-2 region: a duplication model," *Transplant. Proc.* **3:**176–179 (1971).
177. H. Meier, D.D. Myers and R.J. Huebner, "Genetic control by the hr-locus of susceptibility and resistance to leukemia," *Proc. Natl. Acad. Sci. U.S.A.* **63:**759–766 (1969).
178. A. Calvin and J.F. Fries, "Striking prevalence of ankylosing spondylitis in 'healthy' W27 positive males and females," *N. Engl. J. Med.* **293:**835–839 (1975).
179. F.C. Arnett, Jr., "The implications of HL-A W27," *Ann. Intern. Med.* (Editorial) **84:**94–95 (1976).
180. L.M. Cohen, K.K. Mittal, F.R. Schmid, L.F. Rogers and K.L. Cohen, "Increased risk for spondylitic stigmata in apparently healthy HL-A W27 men," *Ann. Intern. Med.* **84:**1–7 (1976).
181. J.L. Amiel, "Study of the leukocyte phenotypes in Hodgkin's dis-

ease," in *Histocompatibility Testing,* pp. 79–81. (E.S. Curtoni, P.L. Matting, and M.R. Tosi, eds.) Copenhagen: Munksgaard, 1967.

182. J.F. Forbes and P.J. Morris, "Leukocyte antigens in Hodgkin's disease," *Lancet* **2:**849–851 (1970).
183. J.D. Zervas, I.W. Delamore and M.C.G. Israels, "Leucocyte phenotype in Hodgkin's disease," *Lancet* **2:**634–635 (1970).
184. W.F. Bodmer, "Genetic factors in Hodgkin's disease: association with a disease-susceptibility locus (DSA) in the HL-A region," *Natl. Cancer Inst. Monogr.* **36:**127–134 (1973).
185. A. Svejgaard, P. Platz, L.P. Ryder, L.S. Nielsen and M. Thomsen, "HL-A and disease associations: a survey," *Transplant Rev.* **22:**3–43 (1975).
186. J.A. Falk and D. Osoba, "The association of the human histocompatibility system with Hodgkin's disease," *J. Immunogenet.* **1:**53–61 (1974).
187. F. Kissmeyer-Nielsen, L.U. Lamm, K.E. Kjerbye, K. Bjørn Jensen, A.M. Nordentoft, K. Thorling and J. Hastrup, "HL-A phenotypes in Hodgkin's disease," in *Histocompatibility Testing,* pp. 593–599, (J. Dausset and J. Colombani, eds.) Copenhagen: Munksgaard, 1972.
188. J. Falk and D. Osoba, "The HLA system and survival in malignant disease: Hodgkin's disease and carcinoma of the breast," in *HLA and Malignancy. Progress in Clinical and Biological Research,* Vol. 16, pp. 205–216, (G.P. Murphy, E. Cohen, J.E. Fitzpatrick and D. Pressman, eds). New York: A.R. Liss, 1977.
189. F. Kissmeyer-Nielsen, K.E Kjerbye and L.U. Lamm, "HL-A in Hodgkin's disease. III. A prospective study," *Transplant. Rev.* **22:**168–174 (1975).
190. J.A. Hansen, C.W. Young, C. Whitsett, D.C. Case, Jr., C. Jersild, R.A. Good and B. Dupont, "HL-A and MLC typing in patients with Hodgkin's disease," in *HLA and Malignancy. Progress in Clinical and Biological Research,* Vol. 16, pp. 217–227, (G.P. Murphy, E. Cohen, J.E. Fitzpatrick and D. Pressman, eds.) New York: A.R. Liss, 1977.
191. R. Harris, S.D. Lawler and R.T.D. Oliver, "The HLA system in acute leukaemia and Hodgkin's disease," *Br. Med. Bull.* **34:**301–304 (1978).
192. G.N. Rogentine, R.J. Trapani, R.A. Yankee and E.S. Henderson, "HL-A antigens and acute lymphocytic leukemia: the nature of the HL-A 2 association," *Tissue Antigens* **3:**470–476 (1973).
193. A.R. Sanderson, G. Hossein Mahour, N. Jaffe and L. Das, "Incidence of HL-A in acute lymphocytic leukemia," *Transplantation* **16:**672–674 (1973).

194. E. Gluckman, F. Lemarchand, A. Nunez-Roldan, J. Hors and J. Dausset, "Possible excess of HLA-A2 homozygous among aplastic anemia and acute lymphoblastic leukemia (ALL)," in *HLA and Disease: Predisposition to Disease and Clinical Implications,* p.226. Paris: INSERM, 1976.
195. M. Jeannet and C. Magnin, "HLA antigens in malignant diseases," *Transplant. Proc.* **3**:1301–1303 (1971).
196. R. Harris, S.R. Zuhrie, C.B. Freeman, G.M. Taylor, J.E. MacIver, C.G. Geary, I.W. Delamore, P.J. Hull and J.A. Tooth, "Active immunotherapy in acute myelogenous leukaemia and the induction of second and subsequent remissions," *Br. J. Cancer* **37**:282–288 (1978).
197. F. Dick, I. Fortuny, A. Theologides, J. Greally, N. Wood and E.J. Yunis, "HL-A and lymphoid tumors," *Cancer Res.* **32**:2608–2611 (1972).
198. F.R. Davey, J.B. Henry and A.J. Gottlieb, "HL-A antigens and acute lymphocytic leukemia," *Am. J. Clin. Pathol.* **61**:662–665 (1974).
199. P.T. Klouda, S.D. Lawler, M.M. Till and R.M. Hardisty, "Acute lymphoblastic leukaemia and HL-A: a prospective study," *Tissue Antigens* **4**:262–265 (1974).
200. J.P. Hester, R. Rossen, J. Templeton, K. McCredie and E. Freireich, "Histocompatibility antigens in adult acute leukemia," in *HLA and Malignancy. Progress in Clinical and Biological Research,* Vol. 16, pp. 71–79, (G.P. Murphy, E. Cohen, J.E. Fitzpatrick, and D. Pressman, eds.) New York: A.R. Liss, 1977.
201. J.R. Batchelor, J.H. Edwards and J. Stuart, Letter to the Editor, *Lancet* **1**:699 (1971).
202. E. Cohen, D.P. Singal, U. Khurana, S.G. Gregory, C. Cox, L. Sinks, E. Henderson, J.E. Fitzpatrick and P. Higby, "HLA-A9 and survival in acute lymphocytic leukemia and myelocytic leukemia," in *HLA and Malignancy. Progress in Clinical and Biological Research,* Vol. 16, pp. 65–70, (G.P. Murphy, E. Cohen, J.E. Fitzpatrick, and D. Pressman, eds.) New York: A.R. Liss, 1977.
203. R.P. Warren, R. Storb, D.D. Nguyen and E.D. Thomas, "Association between leucocyte group 5a antigen and acute lymphoblastic leukemia," *Lancet* **1**:509–510 (1977).
204. R.P. Warren, R. Storb, N.J. Nelson, G.E. Sale and E.D. Thomas, "Increased frequency of the group 5a antigen in patients with hematologic malignancies," *Tissue Antigens* **18**:85–91 (1981).
205. J.M. MacSween, L.A. Fernandez, S.L. Eastwood and A.F. Pyesmany, "Restricted genetic heterogeneity in families of patients with acute lymphocytic leukemia," *Tissue Antigens* **16**:70–72 (1980).

206. C. Werner-Favre and M. Jeannet, "HLA compatibility in couples with children suffering from acute leukemia or aplastic anemia," *Tissue Antigens* **13:**307–309 (1979).

207. M.S. Pollack and D. DuBois, "Non-HLA antibodies in common typing sera: effects on HLA antigen frequency data," in *HLA and Disease: Predisposition to Disease and Clinical Implications,* First International Symposium on HLA and Diseases, p. 232, Paris: INSERM, 1976.

208. R.L. Walford, E. Zeller, L. Combs and P. Konrad, "HLA specificities in acute and chronic lymphatic leukemia," *Transplant. Proc.* **3:**1297–1300 (1971).

209. J.J. Marchalonis, P.J. Morris and A.W. Harris, "Speculations on the function of immune response genes," *J. Immunogenet.* **1:**63–77 (1974).

210. L. Schlosstein, P.I. Terasaki, R. Bluestone et al., "High association of an HL-A antigen, W27, with ankylosing spondylitis," *N. Engl. J. Med.* **288:**704–706 (1973).

211. W.B. Bias and G.A. Chase, "Genetic implications of HLA and diseases associations," *Transplant. Proc.* **9:**531–542 (1977).

212. G.D. Pegrum, I.C. Balfour and C.A. Evans, "HLA antigens on leukemic cells," *Br. J. Haematol.* **19:**493–498 (1970).

213. M. Jeannet and C. Magnin, "HLA antigens in malignant disease," *Transplant Proc.* **3:**1301–1303 (1971).

214. J. Wentzel, A.S. Carvalho and G.M. Taylor, "HL-A and acute myeloid leukemia," in *Histocompatibility Testing,* pp. 813–823, (F. Kissmeyer-Nielsen, ed.) Copenhagen: Munksgaard, 1975.

215. M. Benbunan, M. Sapliery, A. Owens, N. Husson, A. Bussel, M. Reboul, H. Dastot and E. Czazar, "Acute myelocytic leukemia, chronic myeloid leukemia, and HLA antigens," in *HLA and Disease: Predisposition to Disease and Clinical Implications,* p. 217, Paris: INSERM, 1976.

216. R.T.D. Oliver, P. Klouda and S. Lauder, "HLA associated resistance factors and myelogenous leukemia," in *HLA and Disease: Predisposition to Disease and Clinical Implications,* First International Symposium on HLA and Diseases, p. 231, Paris: INSERM, 1976.

217. E.D. Albert, B. Nisperos and E.D. Thomas, "HLA antigens and haplotypes in acute leukemia," *Leukemia Res.* **1:**261–269 (1977).

218. E. Heise, E. Parrish and R. Cooper, "HLA-B17 and the HLA-A1, B17 haplotype in acute myelogenous leukemia," *Tissue Antigens* **14:**98–104 (1979).

219. R.T.D. Oliver, A. Pillai, P.T. Klouda and S.D. Lawler, "HLA linked resistance factors and survival in acute myelogenous leukemia," *Cancer* **39:**2337–2341 (1977).

220. J.P. Hester, R. Rossen and J. Trujillo, "Frequency of HLA antigens in chronic myelocytic leukemia," *South. Med. J.* **70:**691–693 (1977).
221. I.A. Uchida and H.C. Soltan, "Evaluation of dermatoglyphics in medical genetics," *Pediatr. Clin. N. Am.* **10:**409–422 (1963).
222. R. Achs, R. G. Harper and M. Siegel, "Unusual dermatoglyphic findings associated with rubella embryopathy," *N. Engl. J. Med.* **274:**148–150 (1966).
223. M. Alter and R.J. Schulenberg, "Dermatoglyphics in the rubella syndrome," *J. Am. Med. Assoc.* **197:**685–688 (1966).
224. P.A. Davies and V. Smallpiece, "The single transverse palmar crease in infants and children," *Dev. Med. Child Neurol.* **5:**491–496 (1963).
225. J. Aleksandrowicz, Z. Schiffer, and T. Debski, "Dermatoglyphics in leukemia," *Lancet* (Letter to the Editor) **2:**1364 (1966).
226. M.A. Menser and S.G. Purvis-Smith, "Dermatoglyphic defects in children with leukemia," *Lancet* **1:**1076–1078 (1969).
227. F. Rosner, "Dermatoglyphics of leukemic children," *Lancet* (Letter to the Editor) **1:**272–273 (1969).
228. J.L. Verbov, "Dermatoglyphics and leukemia," *Lancet* (Letter to the Editor) **2:**323 (1969).
229. W. Wertelecki, C.C. Plato and J. F. Fraumeni, Jr., "Dermatoglyphics in leukemia," *Lancet* (Letter to the Editor) **2:**806–807(1969).
230. A.H. Nora, J.J. Nora and D.J. Fernback, "Dermatoglyphics and leukemia," *Lancet* (Letter to the Editor) **2:**905–906 (1969).
231. M. Zahalkova and M. Belusa, "Dermatoglyphics in children with leukemia," *Lancet* **1:**1236 (1970).
232. R. Shirley and R.G. DeSai, "Association of leukemia and blood groups," *J. Med. Genet.* **2:**189–191 (1965).
233. F. Vogel, "ABO blood groups and disease," *Am. J. Hum. Genet.* **22:**464–475 (1970).
234. A.L. Lather and A.H. Zaki, "Clinical uses of starch gel electrophoresis with special reference to leukemia," *Clin. Chim. Acta* **5:**22–25 (1960).
235. A.C. Peacock, "Serum haptoglobin type and leukemia: an association with possible etiological significance," *J. Natl. Cancer Inst.* **36:**631–639 (1966).
236. G.G. Wendt, J. Krüger and I. Kindermann, "Serum gruppen und Krankheit," *Humangenetik* **6:**281–299 (1968).
237. F. Galatius-Jensen, "The use of serum haptoglobin patterns in cases of disputed paternity," *Meth. Forensic Sci.* **1:**497–538 (1962).
238. M. Gurda, "Die Korrelation zwischen den Gc-Typ und den ver schiedenen Leukämie formen," *Folia Haematol.* **91:**452–456 (1969).
239. T. Han, H. Ozer, N. Sadamori, L. Emrich, G.A. Gomez, E.S. Hen-

derson, M.L. Bloom and A.A. Sandberg, "Prognostic importance of cytogenetic abnormalities in patients with chronic lymphocytic leukemia," *N. Eng. J. Med.* **310:**288–292 (1984).

240. P.E. Crossen, "Giemsa banding patterns in chronic lymphocytic leukaemia," *Humangenetik* **27:**151–156 (1975).
241. F. Mitelman and G. Levan, "Clustering of aberrations to specific chromosomes in human neoplasms. III. Incidence and geographic distribution of chromosome aberration in 856 cases," *Hereditas* **89:**207–232 (1978).
242. M. Morita, J. Minowada and A.A. Sandberg, "Chromosomes and causation of human cancer and leukemia. XLV. Chromosome patterns in stimulated lymphocytes of chronic lymphocytic leukemia," *Cancer Genet. Cytogenet.* **3:**293–306 (1981).
243. P. Nowell, T.V. Shankey, J. Finan, D. Guerry and E. Besa, "Proliferation, differentiation, and cytogenetics of chronic leukemic B lymphocytes cultured with mitomycin-treated normal cells," *Blood* **57:**444–451 (1981).
244. K. Autio, O. Turunen, O. Penttilä, E. Erämaa, A. de la Chapelle and J. Schröder, "Human chronic lymphocytic leukemia: karyotypes in different lymphocyte populations," *Cancer Genet. Cytogenet.* **1:**147–155 (1979).
245. J. Schröder, P. Vuopio and K. Autio, "Chromosome changes in human chronic lymphocytic leukemia," *Cancer Genet. Cytogenet.* **4:**11–21 (1981).
246. K-H. Robert, G. Gahrton, K. Friberg, L. Zech and B. Nilsson, "Extra chromosome 12 and prognosis in chronic lymphocytic leukaemia. *Scand. J. Haematol.* **28:**163–168 (1982).

6 Physical and Chemical Agents

IONIZING RADIATION

Ionizing radiation consists mainly of electromagnetic waves that remove charges from or add them to electrically neutral atoms or molecules. There are a number of different kinds of ionizing radiation including x rays and gamma rays (photons), neutrons (uncharged particles), beta particles (also termed electrons), and alpha particles (which are helium nuclei). Effects on biological systems depend upon the energy absorption per unit of tissue (measured in rads) and upon the rate of deposition of energy in tissue (linear energy transfer or LET). As the dose decreases, low-LET radiations become less effective per rad whereas high-LET radiations do not. Since the various types of ionizing radiation display differences in relative biologic effectiveness (RBE) there is a need for a measure of dose-equivalence. The rem is defined as the absorbed dose in rads multiplied by a function of relative biologic effectiveness depending upon the specific type of radiation and the particular type of effect at that dose. This standardization is necessary because exposures may consist of a combination of high-LET and low-LET radiation [1, 2].

METHODOLOGIC ISSUES

Carcinogenic effects from radiation are believed to be a function of dose but precise mathematical relationships are not certain on the basis of human data. The threshold concept (the idea that there may be a threshold dose below which radiation is not carcinogenic) does not appear to be supported by animal or human evidence, including the experience of the Japanese A-bomb survivors. A number of different dose-response models have been proposed including a linear, a quadratic, a linear-quadratic with a linear component at low doses, and a quadratic dose-response pattern for high doses, but neither radiobiologic theory nor human data appear to be sufficient to establish a dose-response model that can be used with complete confidence for any specific type of cancer. Human data are particularly scarce in the low dose region and risk estimates are quite unstable [1, 2].

Another related area of great uncertainty is the mechanism of action by which somatic cell carcinogenesis occurs. Although it is thought that DNA may be involved, the evidence supporting the concept that mutation occurs is considered somewhat equivocal. Target theory and the theory of dual radiation action are two related hypothetical constructs in radiation biology that provide the theoretical basis for the dose-response models proposed. Although experimental data support these theories, neither has been unequivocably proven [1, 2].

Data from animal models suggest that the effects of low-LET radiation depends both on the magnitude of the absorbed dose and on its dose-rate or temporal distribution. This is particularly relevant in many occupational exposures in humans. Evidence based on human data is insufficient to conclude whether there is a dose-rate reduction factor (a postulated decreased effectiveness for doses given in a continuous or fractionated fashion over a long interval in time) because of the hypothesis, strongly supported by animal data, that a given sublesion produced is more likely to have healed before another occurs at the same or adjacent location [1, 2].

Another major methodologic problem involves the calculation of dose that may be based on an inappropriate combination of high-LET and low-LET radiation without adjustment for differences in RBE. Also, studies in which rad was used as the measure of absorbed dose may be inappropriately compared with studies utilizing rad-kerma (1 rad-kerma being equivalent to approximately 0.5 rad in whole body exposure to gamma rays and to 0.2 rad in whole body exposure to neutrons, often not specified as the unit of measure) [1, 2].

The effects of age at exposure may vary between individuals in a given population, between subgroups within a population and between populations. The latency period may also vary considerably between individuals even when differences in age at exposure, sex, and dose are adjusted for. Latency, including the minimal latency period, also varies for different types of cancer. Very little is known about the association between radiation effects and other host factors such as genetic characteristics, immune competence, hormonal status, capacity for DNA repair, and sociodemographic and lifestyle factors [1, 2].

Two models have been used to predict the rate of radiogenic cancer following exposure. The relative risk model is based on the assumption that radiogenic risk is proportional to the age-specific incidence rate of spontaneous cancer for a specific cancer type, with excess risk increasing as the spontaneous incidence increases. The absolute risk model postulates that excess risk is constant. Neither model appears to fit all situations [1, 2].

Studies examining the relationship between radiation and leukemia are often complicated by small numbers in cell-specific subgroups and in individual-dose ranges the numbers are even smaller. Given this constraint, any estimate of a leukemia induction rate among those exposed to a specific dose of radiation is subject to substantial random error. For these reasons, the results of the studies cannot be used to definitely select one or another of the proposed dose-response models as the "true" model [1, 2].

The carcinogenic effects of ionizing radiation is a burgeoning area of study in epidemiology and other scientific disciplines. Even though ionizing radiation has probably received more attention than most other environmental exposures, knowledge of its effects upon human health, including carcinogenesis, is incomplete. Theories about (and postulated models of) dose-response, latency, mechanisms of somatic cell carcinogenesis, and projection models of estimated risk are constantly undergoing scrutiny with further refinement, development of new models, and rejection of postulates (such as the threshold concept) occurring with great rapidity.

RADIATION AND LEUKEMOGENESIS

Though there is an extensive literature on the relationship between radiation and leukemogenesis, the mechanisms and threshold levels, if they exist, are not well known [3, 4]. Acute forms of leukemia (including acute lymphocytic leukemia and the subtypes of acute nonlymphocytic leukemia) and chronic myelocytic leukemia have been found to be induced by radiation; there is no evidence of an association between CLL and radiation [1].

JAPANESE ATOMIC BOMB SURVIVOR STUDIES

A major source of information on the effects of radiation is from the Japanese atomic bomb survivor studies. Long-term follow-up studies of the large population of Japanese survivors of the atomic bombs detonated in 1945 in Hiroshima and Nagasaki have provided investigators with a unique opportunity to improve understanding of mechanisms of carcinogenesis. Although most workers agree that understanding of health effects of low-dose radiation cannot be completely resolved by human studies, epidemiologic data such as that produced by ongoing research at the Radiation Effects Research Foundation of the Japanese A-bomb survivors is a necessary complement of experimental work directed at further delineation of the mechanisms by which ionizing radiation produces biologic damage [5].

Since 1950 an extended sample of 109,000 subjects, of whom 82,000 were A-bomb survivors and 27,000 were not in the city at the time of the bomb (ATB), has been followed in great detail in order to determine incidence and mortality rates for various conditions (including leukemia) in A-bomb survivors compared with those not in the city ATB. This sample, known as the Life Span Study (LSS) Sample, has been characterized in terms of age, sex, and dose of radiation received ATB. The population from which the 82,000 survivors in the LSS Sample were selected consisted of 117,000 survivors. Attempts were made to ascertain leukemia mortality in the additional 35,000 survivors not included in the LSS but difficulties in monitoring the group and the scattered residence of a significant number of its members made this effort unfeasible [5, 6].

Kato and Schull, in the most recent follow-up of cancer mortality through 1978, noted that the greatest proportion of excess leukemia mortality (83%) attributed to radiation from the atomic bombs occurred in the interval 5 to 21 years following exposure and the remainder (17%) 22 to 33 years following the exposure. The greatest peak in leukemia mortality occurred in 1950–54. Following that time period the excess began to decrease; by 1975–78 an increased mortality with dose was observed only in Hiroshima [7].

Ichimaru et al. examined the incidence of leukemia in the LSS from 1950 to 1971 by radiation dose, years after exposure, age, and type of leukemia [8]. By June 1972, 1,559 cases of definite and probable leukemia had been registered in the Leukemia Registry (maintained by the Hematology Departments of the Hiroshima and Nagasaki Schools of Medicine, and the Radiation Effects Research Foundation Departments of Medicine, Pathology, Epidemiology, and Statistics); of the 1,559 cases, 422 were A-bomb survivors. This registry is believed to provide almost complete ascertainment of leukemia cases with onset in Hiroshima and Nagasaki since the time of the bombing [8]. Additional follow-up through 1978 has more recently been reported for Nagasaki by Wakabayashi et al. [9]. These investigators estimated that the excess incidence of leukemia, all types combined, was 42% for the period 1959–78; this excess decreased with time and reached zero by 1981.

The radiation produced by the bomb exploded in Nagasaki was primarily pure gamma ray while the Hiroshima bomb also had a small neutron ray component in addition to the predominant gamma ray exposure. Early estimates suggested that neutrons constituted up to 25% of the radiation produced by the Hiroshima bomb [10] but more recently a number of investigators have indicated that a major reduction in the neutron component estimate may be required [11–13].

Examination of the distribution of leukemia cases by cell type has shown differences by dose and by city. Older estimates of dose, now undergoing reevaluation, defined a T65 "total dose" as the simple sum of gamma and neutron doses. In addition to the current controversy over the estimated neutron component, a number of other factors, not originally adjusted for in determination of the T65 dose, have also begun to receive attention because of the differences noted in cell-type distributions between the two cities. (Such factors include sociodemographic variables, differences in housing, shielding differences, and a number of topographic characteristics.) Those in Hiroshima at the time of the bomb who developed leukemia were more likely to develop CML (33.7% of leukemics who had 100+ rads of exposure; 50% of leukemics who had exposures of 1–99 rads) than those in Nagasaki (CML representing 16.3% of those leukemics with 100+ rads of exposure; and 25% of leukemics exposed to radiation levels of 1–99 rads). The excess CML developing in Hiroshima was offset by a deficit of cases of ANLL (31.3% of all leukemias for exposures of 100+ rads; 26.3% for exposures of 1–99 rads) compared with Nagasaki (where ANLL comprised 51.1% of all leukemias for 100+ rads of exposure; 54.2% for 1–99 rads). The proportion with ALL was approximately 34% for those leukemics exposed to 100+ rads and approximately 22% for those exposed to 1–99 rads in both Hiroshima and Nagasaki (see Tables 6-1 and 6-2). (Ichimaru et al. noted that ten cases of CLL were registered as A-bomb survivors in Nagasaki but none of them received as much as one rad [14]. No case of CLL was registered among those known to have received one or more rads in either Hiroshima or Nagasaki.)

Ichimaru et al. further noted that examination of the distribution of the Leukemia Registry cases appeared to show that the peak years of onset for cases receiving 100 or more rads were 1951–53 [8]. However, they also noted that the number of leukemia cases who had not received significant doses of radiation or were not in Hiroshima or Nagasaki ATB actually increased in the years following exposure even though the population in these cities remained fairly stable. This was attributed to better and more complete leukemia case ascertainment in the cities and areas adjacent to the cities. It is somewhat difficult to interpret the meaning of the increase in the proportion of leukemics who had no radiation exposure or extremely low levels of exposure during the later 1950's and 1960's. Presumably it is due to better ascertainment but other types of nonidentified leukemogenic exposures may be playing a significant role.

Table 6-3 contains a comparison of crude annual leukemia incidence rates among those in the LSS sample who received 100 or more rads in

Table 6-1 Distribution of definite and probable leukemia cases in the leukemia registry by type and dose – Nagasaki

	Exposed (T65 dose in rads)								Nonexposed (born after A-bomb)	
	Unknown		100+		1–99		<1			
Type of leukemia	No.	%	No.	%	No.	%	No.	%	No.	%
ANLL (includes AGL and acute monocytic leukemia)	4	36.4	22	51.1	13	54.2	46	52.3	109	45.6
ALL	2	18.2	9	20.9	3	12.5	10	11.4	98	41.0
Undifferentiated and other (acute)	2	18.2	5	11.7	2	8.3	12	13.6	20	8.4
CML	3	27.2	7	16.3	6	25.0	11	12.5	12	5.0
CLL	0	—	0	—	0	—	9	10.2	0	—
Total (all leukemia types combined)	11	100.0	43	100.0	24	100.0	88	100.0	239	100.0

Source: Ichimaru et al. (1976) [14].

Table 6-2 Distribution of definite and probable leukemia cases in the leukemia registry by type and dose – Hiroshima

	Exposed (T65 dose in rads)								Nonexposed (born after A-bomb)	
	Unknown		100+		1–99		< 1			
Type of leukemia	No.	%	No.	%	No.	%	No.	%	No.	%
ANLL (includes AGL and acute monocytic leukemia)	2	12.3	26	31.3	20	26.3	40	49.4	90	43.5
ALL	2	12.5	18	21.8	9	11.8	12	14.8	55	26.6
Undifferentiated and other (acute)	5	31.4	11	13.2	9	11.9	14	17.3	50	24.2
CML	7	43.8	28	33.7	38	50.0	15	18.5	12	5.7
CLL	0	—	0	—	0	—	0	—	0	—
Total (all leukemia types combined)	16	100.0	83	100.0	76	100.0	81	100.0	207	100.0

Source: Ichimaru et al. (1976) [14].

Table 6-3 Comparison of crude annual incidence per 100,000 among those in the extended Life Span Study Sample who received 100 or more rads in Hiroshima and Nagasaki by dose, age ATB, and type of leukemia

	Age at the time of the bomb				
Type of leukemia	< 15	15–29	30–44	45+	Total
October 1950–September 1955					
Person-years	7597	10 121	6489	5311	29 518
AGL	0.0 (0)[a]	39.5 (4)	30.8 (2)	00.0 (0)	20.3 (6)
ALL	52.7 (4)	29.6 (3)	0.0 (0)	0.0 (0)	23.7 (7)
CGL	65.9 (5)	29.7 (3)	46.2 (3)	0.0 (0)	37.3 (11)
October 1955–September 1960					
Person-years	7476	9964	6225	4496	28 161
AGL	13.4 (1)	0.0 (0)	64.3 (4)	44.5 (2)	24.9 (7)
ALL	13.4 (1)	10.0 (1)	0.0 (0)	0.0 (0)	7.1 (2)
CGL	0.0 (0)	0.0 (0)	0.0 (0)	22.2 (1)	3.6 (1)
October 1960–December 1971					
Person-years	16 569	21 926	12 866	6985	58 346
AGL	0.0 (0)	4.6 (1)	15.5 (2)	43.0 (3)	8.6 (5)
ALL	0.0 (0)	4.6 (1)	15.5 (2)	14.3 (1)	6.9 (4)
CGL	0.0 (0)	4.6 (1)	7.8 (1)	0.0 (0)	3.4 (2)

[a]Number of cases in parentheses.
Source: Ichimaru et al. (1976) [14].

Table 6-4 Comparison of relative risks[a] for leukemia incidence by age and dose for Hiroshima and Nagasaki, males and females combined

Dose of radiation (rads)	Age[b]					
	0–9	10–19	20–34	35–49	50+	All ages
0–9	1.000 (7)	1.000 (6)	1.000 (11)	1.000 (23)	1.000 (7)	1.000 (54)
10–49	1.425 (3)	5.407 (7)	1.097 (3)	0.625 (4)	1.581 (3)	1.405 (20)
50–99	2.027 (1)	2.520 (1)	1.294 (1)	1.080 (2)	3.762 (2)	1.770 (7)
100–199	12.443 (4)	9.523 (4)	8.254 (5)	0.687 (1)	6.483 (2)	5.215 (16)
200–299	21.234 (3)	14.001 (3)	13.201 (4)	5.030 (3)	16.955 (2)	10.804 (15)
300–399	15.592 (1)	10.571 (1)	36.582 (5)	10.575 (3)	36.588 (2)	18.902 (12)
400 +	66.005 (7)	25.997 (3)	16.215 (3)	13.111 (5)	26.818 (2)	23.131 (20)

[a]Relative risks were calculated using those who had received 0–9 rads as the comparison group.
[b]Numbers in parentheses represent numbers in each age and dose group.
Source: Beebe et al. (1977) [15].

Hiroshima and Nagasaki, by year of diagnosis of leukemia, age at the time of the bomb, and type of leukemia.

Using those who had received 0–9 rads as the reference group, relative risks for each age group were substantially greater as the dose of radiation increased (Table 6-4). In general, risks were highest for those in the first two decades who had been exposed to 400+ rads and those over age 50 exposed to 300+ rads. For those receiving 10–99 rads exposure, the risk of leukemia was highest in those ages 10–19 [8, 14, 15].

The extensive analyses by Ichimaru et al. [6, 14] confirmed a definite, strong association between leukemia and high-dose radiation, a relatively short latency period between exposure and disease onset (when compared with other cancers associated with radiation exposure), and an increased sensitivity to radiation carcinogenesis in those in the first two decades of life ATB and those over age 50 ATB, compared with those ages 20–49 at the time of exposure.

Jablon and Kato found that those exposed *in utero* did not appear to have experienced an excess risk of leukemia in a 1960's follow-up study of 1,292 children, exposed prenatally in 1945 to the atomic bombs. No excess mortality from leukemia or other cancers was found among the 325 children whose mothers (when pregnant with these children) were within 1,500 meters of the hypocenters although this number (325) was somewhat small and may have been insufficient to detect a significant excess [16].

Other Studies of Leukemia in Persons Exposed to Radiation Fallout

The excess mortality from leukemia (and other cancers) experienced by the Japanese survivors of the atomic bombs described above was primarily due to a single high dose of radiation. Another type of exposure produced by detonation of atomic bombs consists of a low level of ionizing radiation from radioactive fallout at a site removed from the nuclear blast. While most studies examining the association between low-dose radiation from fallout and human cancers have been negative [17–19], Conard et al. noted an excess of thyroid cancer in natives of the Marshall Islands following the atomic-bomb test explosion in 1954 [20].

From 1951 to 1958, southern and eastern areas of Utah were exposed to fallout from 26 nuclear tests. A somewhat controversial study by Lyon et al. [21] reviewed all deaths from childhood (under 15 years of age) cancers occurring in Utah between 1955 and 1975. The investigators assigned cancer deaths to a cohort of either high or low exposure depending upon their age during the time of the fallout and further classified the children by residence in high- versus low-fallout counties. Findings indicated that leukemia mortality among the low-exposure cohort (children under 15 during 1944–50 and those born after 1958) in the high-fallout counties, which were highly rural, was about half that of the population in the U.S. and the remainder of Utah (both of which have higher proportions of urban residents). It is possible that this unexplained observation may be a function of the finding that rural populations have lower overall rates of acute childhood leukemia than do urban populations [22]. Leukemia mortality was 2.44-fold higher in the high-exposure cohort (children under 15 during 1951–58) residing in high-fallout counties compared with mortality in the low-exposure cohort residing in the same counties. Leukemia mortality (1959–67) was highest in children dying at 10–14 years of age who had lived in high-fallout counties. There was also an excess of leukemia in the high-exposure cohort living in low-fallout counties which had an SMR of 1.28 that was not statistically significant but in the same direction as that for the high-fallout counties.

Lyon et al. point out certain limitations of their study including their inability to determine the histologic type of leukemia from the death certificate, possible misclassification of cause of death on the certificate, absence of identification of longest place of residence as opposed to last place of residence which is listed on the certificate, absence of data on actual, individual radiation exposure, an absence of ascertainment of other potential etiologic factors, i.e., heavier than usual exposure of mothers of cases, while pregnant, to diagnostic x rays [21, 23].

Other critiques have noted that the correlations are ecologic because the actual radiation exposures experienced by individuals are unknown. Mortality from childhood cancers other than leukemia was not higher in high-fallout counties when compared with low-fallout counties during 1955–75. Therefore, the observed association between leukemia and apparent exposure to radiation fallout may have been artifactual, or it may have been a normal fluctuation that may occasionally occur with rare cancers in small populations [24].

The Comptroller General's GAO Report to Congress also noted that Lyon et al. did not allow for a two-year minimum latency period between radiation exposure and onset of leukemia since exposure occurred in 1951–58 and leukemia mortality was tabulated beginning in 1955 [25]. The Lyon analysis compared observed leukemia deaths among 10–14 year olds in the high-exposure cohort to a very low number of deaths for that age group in the low-exposure cohort. (There were no leukemia deaths among boys 10–14 years old between 1944 and 1953 in the low-exposure cohort.) This may have resulted in an apparent overestimate of the excess deaths associated with fallout exposures since the baseline comparison levels were substantially lower than the U.S. average. The GAO reanalysis of the Lyon et al. data resulted in a lower but still significant estimated excess for the high-exposure cohort in the high-exposure counties and a much lower estimated excess for that cohort in the low-exposure counties [25].

The "elevated" leukemia mortality rate for the high-exposure cohort (in the high-exposure counties when compared to the rate in the low-exposure cohort in low-exposure counties) was not higher than the U.S. average age-specific leukemia rates. Enstrom has pointed out that there is no conclusive epidemiologic evidence for radiation-induced childhood leukemia involving external doses less than 100 rads [26]. He argued that earlier investigators found no evidence of radiation-related disease in Utah [17–19], that there was some possibility of misdiagnosis of childhood leukemia in the high-exposure counties in southern Utah (with fewer board-certified physicians than in the low-exposure counties in northern Utah), that the addition of only two more leukemia deaths to the reported 17 that occurred in the low-exposure cohort would remove the significance of the association between the exposure cohort and leukemia, and that the choice of geographic boundaries and time periods by Lyon et al. was somewhat artificial.

Two other studies also cast doubt upon the findings of Lyon et al. Beck and Krey, using recent measurements of residual cesium-137 and plutonium in soil from 150 sites in 56 population centers in Utah and western Colorado, found that the residents of the populous northern "low-fallout" counties of northern Utah actually received a higher mean dose and a significantly greater population dose than did the residents of most counties closer to the test site. However, the doses calculated for all areas of Utah were much lower than that currently believed to be required for a significant increase in leukemia mortality and "far too low to account for the excess childhood leukemia mortality reported by Lyon et al." [27]. Land et al. used childhood cancer mortality data from the National Center for Health Statistics 1950–78 and examined geographic and temporal variation for leukemia and childhood cancer other than leukemia mortality, comparing northern and southern Utah (using the same geographic partitioning as Lyon et al.), eastern Oregon, Iowa, and the entire U.S. They found no statistically significant geographic or temporal differences, but an apparent downward trend over time in childhood leukemia mortality. They concluded that their data did not support the Lyon et al. conclusion of increased risk associated with exposure to fallout in southern Utah and that, most likely, Lyon et al. had based their conclusion on "an anomalously low leukemia rate in southern Utah during the period 1944 to 1949" [28].

Caldwell et al. showed a significant increase in leukemia (9 cases observed while 3.5 were expected) among 3,224 men who participated in military maneuvers during the 1957 nuclear detonation of a 44-kiloton bomb called "Smoky" in Nevada [27]. Nine cases (4 AML, 3 CML, 1 ALL, and one with hairy cell leukemia) occurred when 3.5 were expected. The only exposure data available were in the form of film badges showing an average whole-body gamma dose of 466.2 mrem for the entire cohort and a mean dose of 1,167 mrem for cases. However, as the authors indicated, there are several problems with the exposure data including: a significant number of badge dose readings of zero, the badge dose representing a cumulative dose received from the entire series of tests with no specific information about the individual doses comprising the cumulative dose that may have come from other sources of radiation exposure; badge doses reflecting only external radiation exposures from gamma and beta radiation; and no information about neutron exposure or internal exposure. There was no monitoring for internal exposures due to ingested or inhaled radioactive material. The cases or their next-of-kin were questioned as to other possible leukemogenic exposures. Three of the cases may have had additional exposures (from pesticides, chloramphenicol, and cosmic radiation in the case of a commerical airline pilot). The latent period between exposure and onset of leukemia averaged 14.4 years, longer than that of the Japanese atomic-bomb survivors for whom minimal latency was approximately four years and maximum latency was 10–12 years. Follow-up of the Smoky cohort was incomplete (76%) and hence conclusions must be interpreted with caution (see Table 6-5).

Leukemia in Radiologists

Seltser and Sartwell [28] and Matanoski et al. [29, 30] examined the mortality experience of American radiologists (members of the Radiological

Society of North America) as compared to that of internists who belonged to the American College of Physicians, and members of the American Academy of Ophthalmology and Otolaryngology. The leukemia death rates (cell types not specified) among radiologists were 8.8 times higher than that of ophthalmologists in the 1920–29 cohort of radiologists, 3.4 times higher than ophthalmologists in the 1930–39 cohort, and similar to ophthalmologists in the 1940–49 cohort. Ophthalmologists were selected as the main comparison group since they were believed to have the lowest occupational exposure to radiation of the three physician comparison groups. In subsequent cohorts, leukemia deaths decreased but lymphoma (especially multiple myeloma) deaths increased beginning with the cohort of radiologists who entered the specialty in 1940–49.

Other studies have also found excessive occurrence of leukemia (cell type not specified) among American radiologists in comparison with other physicians [31, 32]. These reports appear to establish that repeated exposure to fairly small doses of radiation can increase leukemia risk [33, 34].

Two studies of medical personnel working with x-ray equipment did not find an increased incidence of leukemia; one was carried out in British radiologists [35] and one in U.S. Army x-ray technicians [36]. Court-Brown and Doll speculated that radiologists in Britain may have exercised greater precaution than U.S. radiologists and therefore were less exposed. Miller and Jablon postulated that the x-ray technicians were exposed to radiation levels insufficient to produce leukemia (see Table 6-6).

Other Occupational Radiation Exposure

A great deal of controversy has surrounded the findings and methodology employed by Mancuso et al. in their proportional mortality analysis of cancer and noncancer deaths among Hanford Nuclear Facility workers [37]. These investigators analyzed deaths of white males among those who were exposed to radiation and those not exposed and found a higher proportion of "irradiated" workers among the cancer deaths than among the noncancer deaths. Irradiated workers were defined as those whose radiation badges showed a positive reading at any time between 1944 and 1972. Mancuso et al. divided the cases into five age groups and five radiation dose groups and ranked the proportion of cancer victims within the five dose ranges for each age group. Higher radiation doses appeared to correlate with a higher proportion of cancer deaths. They also compared the mean radiation doses of persons dying of cancer with the mean dose of all other noncancer causes of death. For eight categories of malignant neoplasms, the mean doses were greater than that of all deceased workers combined. For all cancers and for three specific types of malignancies including reticuloendothelial malignancies (particularly the subcategories of multiple

Table 6-5 Studies examining the relationship between leukemia and exposure to nuclear weapons, fallout, or nuclear facilities

Author(s), year	Study group	Type of radiation	Amt. of radiation	Findings with respect to leukemia
Jablon et al., 1970 [16] Ichimaru, M. et al., 1976 [14] Ichimaru M. et al., 1978 [8] Beebe et al., 1977 [15] Ishimaru, T. et al., 1979 [6]	Japanese atomic-bomb survivors	Hiroshima: gamma rays and neutrons Nagasaki: gamma rays primarily	0–400+ rads	Leukemia (including ALL, ANLL, and CML, but *not* CLL) was identified as a late mortality effect of nuclear radiation. There was a stronger radiation carcinogenic effect for Hiroshima.
Lyon et al., 1979 [21] Land, 1979 [24] Lyon et al., 1980 [23] Comptroller General's Report, 1981 [25]	Children in Utah under 15 years of age from 1944–75 including 378 children who died from leukemia or other cancers.	Gamma rays from nuclear tests carried out in Nevada during the 1950's.	Unknown	Observed leukemia mortality increased 2.44-fold for the high-exposure cohort of children living in high fallout counties in southern Utah compared with the exposure cohort of children in the low fallout counties. Leukemia cell types not available on most death certificates.

Caldwell et al., 1980 [29]	3224 men who participated in military maneuvers during the 1957 nuclear test explosion "Smoky"	Gamma rays	0–2.98 rems	Statistically significant increase in leukemia incidence; 9 cases were observed while 3.5 were expected. The interval from time of nuclear test to diagnosis ranged from 2 to 19 years. Cell types included 4 AML, 3 CML, and 1 ALL.
Mancuso et al., 1977 [39] Anderson et al., 1978 [44] Hutchison et al., 1978 [42] Gilbert et al., 1979 [43]	27,960 Hanford Nuclear Facility workers	Gamma rays	0–50 rads	No excess of leukemia mortality noted. However, mortality rates of pancreatic cancer, multiple myeloma, and possibly lung cancer were elevated.
Najarian et al., 1978 [45] Colton et al., 1980 [46]	1722 nuclear workers at the Portsmouth Naval Shipyard. Reanalysis based on 1038 deaths.	Gamma rays	For exposed workers the total cumulative recorded exposure ranged from 0.001 to 64.9 rem.	Original analysis based on 1722 deaths and recollections of next-of-kin as to whether the worker had been employed maintaining nuclear vessels. Reanalysis based on Navy employment records showed that there was no significant excess of leukemia among either exposed or nonexposed workers.

Table 6-6 Follow-up studies of leukemia in radiologists and radiology technicians

Author(s), year	Study group	Type of radiation	Amt. of radiation	Findings with respect to leukemia
Seltser and Sartwell, 1965 [30] Matanoski et al., 1975 [31,32]	16,339 American radiologists	X rays	Unknown	Radiologists who entered the specialty in 1920–39 showed an excess of leukemia when compared to certain other medical specialists. However, radiologists who entered the specialty in 1940–49 had lower leukemia mortality and increased mortality from lymphomas and multiple myeloma. Leukemia cell type(s) not specified.
Court-Brown and Doll, 1958 [37]	1,377 male British radiologists	X rays	Unknown	An increase in leukemia was found among those men who began to practice radiology before 1921.
Miller and Jablon, 1970 [38]	6,560 white male radiology technicians during World War II	X rays	Average exposure was 50 x-ray film studies and 2 fluoroscopic exams per day. 59% of technicians had worked at least 2 yrs.	No statistically significant increase in leukemia.

myeloma and myeloid leukemia), pancreatic and lung cancers, the mean dose of radiation exposure was significantly higher than for individuals dying of all noncancer causes. Mancuso et al. also attempted to estimate the number of cancers caused by radiation and the "doubling dose," i.e., the amount of radiation necessary to double the risk of developing a given disease. Overall, they estimated that the radiation caused an excess of 26 cancers.

The conclusions reached by Mancuso et al. were criticized by many individuals and groups. One serious methodological problem noted was that the mean cumulative exposures were highly skewed by a few exposures to very high doses. Other major criticism of these analyses can be summarized as follows: (1) the statistical methods used were not appropriate—use of proportional mortality rates, or PMR, and inclusion of individuals whose total dose received represented outliers; (2) the study population may have been too small especially when the various age and exposure subgroups of that population were separately considered; (3) appropriate statistical adjustments for confounding variables may not have been made; (4) the conclusion by Mancuso et al. that sensitivity to cancer induction is increased in individuals exposed between ages 20 and 40 may not be justified; (5) several cancers (pancreatic, lung, multiple myeloma, myeloid leukemia) were associated with radiation exposure by Mancuso et al. whereas in the independent analyses, only multiple myeloma and carcinoma of the pancreas were found to be associated [37, 38].

In another analysis by the Mancuso group, statistical adjustment for a number of factors associated with either radiation, cancer, or both still resulted in a higher proportion of cancers for a higher dose level of exposure compared to the lower proportion of cancers for a lower dose level of exposure [39].

The Stewart, Kneale, Mancuso group addressed the five criticisms summarized above as follows: (1) an SMR analysis could not be carried out because a complete follow-up was impossible due to the "existence of an unknown number of deaths which may never be traced"; (2) the size of the study population was fixed; (3) in a reanalysis, some variables were adjusted for, although smoking (for which data were not available) was not; (4) for those dying of cancer before age 40, the mean dose of radiation was the same for those dying of cancers of nonradiosensitive tissues and those dying of causes other than cancer as for those dying of cancers of radiosensitive tissues; however, those dying after age 40 of cancers of radiosensitive tissues had higher cumulative levels of radiation exposure than the other two groups; (5) mean radiation doses for radiation-related cancers (2.04 rads) were higher than those for cancers not associated with radiation exposure (1.24 rads).

There have been other criticisms of the Mancuso et al. findings and conclusions. Hutchison et al. reanalyzed the Hanford data by age and calendar year of death [40]. For each dose interval, the ratio of observed to expected number of deaths or "standardized proportional mortality ratio" was computed and the significance of the relation between proportional mortality ratio and dose was examined by means of a trend statistic. This reduced the number of cancer sites for which a radiation-dose relationship was observed to two: multiple myeloma and pancreatic cancer. This analysis showed no association between exposure to radiation and leukemia.

Gilbert and Marks also reanalyzed the data in two ways [41]. The mortality of Hanford workers was compared with that of the U.S. population. A marked "healthy worker effect" in the Hanford workers was shown as well as an absence of significantly high standardized mortality ratios for any specific disease categories. In another analytic approach, death rates were related to exposure status without reference to an outside population. Hanford workers with higher radiation exposures were compared with other Hanford workers, similar in most variables except for their level of radiation exposures, categorized into four groups: 0–2, 2–5, 5–15, and 15+ rem. Expected deaths for each exposure group were calculated from the combined experience of the groups using the Mantel–Haenszel method, adjusting for calendar year, age, and variables of interest, such as job category and employment status. A test for association of mortality with levels of radiation exposure showed no correlation for all cancers. Gilbert and Marks also found no association between exposure to radiation and lung cancer and leukemia but did find a statistically signficant test for trend for multiple myeloma and cancer of the pancreas.

Anderson noted that Mancuso et al. did not provide complete information on the different histopathologic types of leukemias, although myeloid and lymphatic leukemias were identified separately [42]. He pointed out that various types of leukemia were classified in a group labeled "RES residue." In addition, in the proportionate morality analyses, observed deaths were classified by Mancuso et al. according to the 8th Revision of the International Classification of Diseases whereas expected deaths were classified by the 7th Revision.

Mortality from leukemia and cancer in shipyard nuclear workers in the Portsmouth Naval Shipyard was examined by Najarian and Colton [43]. Next-of-kin were contacted for 592 of 1,722 deaths found in a death certificate review for 1959–77. All deaths were classified as being among either former nuclear or nonnuclear workers depending on information supplied by next-of-kin. Using U.S. age-specific proportional cancer mortality rates for white males as a standard, the observed-to-expected ratio of leukemia deaths was 5.62 among the 146 former nuclear workers. The six observed leukemia deaths were not further specified as to cell type. A reanalysis of deaths among these shipyard workers using information from

Navy employment records, rather than from next-of-kin showed that the observed deaths from hematological malignancies did not differ significantly from those expected for exposed or for nonexposed workers [44]. However, a proportional mortality analysis examining the dose-response relationship showed that the proportion of hematologic malignancies increased with larger cumulated exposures.

None of the epidemiologic studies of miners exposed to alpha particles (radon and its daughters) have demonstrated a leukemia excess, including studies of U.S. uranium miners [45, 46], Canadian fluorspar miners [47], British iron ore miners [48], Czechoslovakian radium miners [49], and U.S. underground metal miners [50]. A number of these investigations reported an excess of lung cancer. The sample size of the miner populations studied may not have been large enough to show increased risk for the various leukemias compared with control groups. The type of radiation (alpha particles from radon and its daughters) to which these miners were exposed, summarized in Table 6-7, may not be as leukemogenic as other types of radiation [45–50].

Association between Leukemia and Radiation for Diagnostic/Therapeutic Reasons

Many studies have examined leukemia risk from both diagnostic as well as therapeutic radiation exposure. Some of the studies have employed a retrospective approach, examining the history of radiation exposure in individuals with leukemia in comparison with controls. Other investigators have carried out follow-up studies on various groups with known exposure to diagnostic or therapeutic radiation. Often the studies have examined the risk of leukemia from radiation exposure to specific anatomic sites such as uterine exposure from pelvimetry in women who were pregnant, "therapeutic" radiation to the spine in individuals with ankylosing spondylitis, or radiation to the scalp of children as treatment for tinea capitis.

Case-Control Studies Many of the key studies of the leukemia risk from medical radiation are summarized in Tables 6-8 and 6-9. Table 6-8 briefly describes several retrospective studies of associations between leukemia and radiation exposure for medical reasons. Three case-control studies examined the association between radiation exposure for medical diagnostic or therapeutic reasons and subsequent development of leukemia in adults. Persons with acute leukemia and those with CML reported a significant excess of diagnostic and therapeutic x-ray exposure in cases in the two studies with substantial sample sizes, 1,414 in the Tri-State study [51–56] and 467 (590 including children) in the study carried out by Gunz and Atkinson [57], compared with controls. The relationship was noted only

Table 6-7 Studies examining the relationship between radiation exposure and subsequent cancer in miners exposed to alpha particles

Author(s), year	Study group	Type of radiation	Amt. of radiation	Findings with respect to leukemia
Saccomanno et al., 1971 [48] Archer et al., 1973 [47]	U.S. uranium miners who had spent from 1 month to 25 years working underground	Alpha particles	0–3,720 working level months (WLM) of radiation. A WLM is approximately equivalent to 1 rad of alpha radiation exposure in the bronchial epithelium of the lungs	Significant excess of lung cancer, particularly small cell undifferentiated, but leukemia mortality not reported.
Devilliers and Windish, 1964 [49]	Canadian fluorspar miners employed from 1933 through 1961	Alpha particles	Unknown	26/119 deaths due to lung cancer (29 times expected). 2/119 deaths due to leukemia (expected rates not given).
Boyd et al., 1970 [50]	British iron ore miners who died between 1948 and 1967	Alpha particles	Average radon air concentration was 100 picocuries per liter	Significant excess of lung cancer mortality but leukemia mortality not reported.
Pirchan and Sikl, 1932 [51]	Czechoslovakian radium miners who died between 1920 and 1938	Alpha particles	Unknown	9/13 autopsied miners died of lung cancer. 47/60 autopsy cases died of cancer. Of the 47 dying of cancer, 43 had primary respiratory neoplasms. Leukemia mortality not reported.
Wagoner et al., 1963 [52]	U.S. underground metal miners who had worked underground for at least 15 yrs.	Alpha particles	Unknown	Threefold increase in mortality from primary respiratory cancer over expected. Leukemia mortality not reported.

Table 6-8 Studies examining the relationship between diagnostic or therapeutic medical radiation and subsequent development of leukemia – (retrospective)

Author(s), year	Study group	Type of radiation	Amt. of radiation	Findings with respect to leukemia
Graham et al., 1966 [53] Gibson et al., 1968 [54] Gibson et al., 1972 [55] Bross and Natarajan, 1972 [56] Bross et al., 1979 [57] Boice and Land, 1979 [58]	1414 adults and 151 children ages 1–4, diagnosed as having leukemia or dying from leukemia 1959–62 in upstate New York and the metropolitan Baltimore and Minneapolis areas (Case-control Tri-State Leukemia Survey)	Adults: diagnostic x-ray films Children: prenatal and postnatal x-ray diagnostic films	Adults: 0–41 films (average bone marrow dose calculated to be <0.5 rad per film) Children: unknown	Adults: Males with both AML and CML were more likely to have received significant amounts of diagnostic x rays compared with controls. Females with leukemia of all types did not differ from controls with respect to radiation exposure. For males with CML, risk following exposure to 11 or more diagnostic x-ray films was 2.2 times that following exposure to 11 or fewer films. Bross et al. suggested that estimates obtained by extrapolation from high dose levels to low dose levels underestimated the risks. However, Boice and Land point out a number of alternative explanations related to confounding factors, potential sources of bias, possible errors in dose determinations as well as problems with the statistical analysis that shed doubt upon the conclusions drawn by Bross et al.

Table 6-8 (continued)

Author(s), year	Study group	Type of radiation	Amt. of radiation	Findings with respect to leukemia
				Children: An increased risk for leukemia was associated with diagnostic x rays received by mothers up to 10 years before conception (60% excess) and preconception diagnostic x rays in fathers (30% excess). Children exposed to several "pathologic" (risk) factors including reproductive wastage, *in utero* radiation, viral disease contracted by the subject more than 12 months before the onset of leukemia had a significantly greater risk than children exposed to only one factor.
Stewart et al., 1958 [61] Stewart, 1973 [62] Bithell and Stewart, 1975 [63]	8513 children in England and Wales who died of cancer, 1953–67	*In utero* exposure to diagnostic x rays	Unknown	Exposure to x rays *in utero* results in a significant increase in malignant disease particularly leukemia (R.R. for ALL = 1.6; R.R. for AML = 1.6; R.R. for other, unspecified = 1.3) compared with controls. Leukemia may be an unrecognized cause of stillbirths, abortions.

Gunz and Atkinson, 1964 [59] *	590 cases of leukemia comprising 87.3% of all leukemias known to have occurred in New Zealand (1958–61)	Diagnostic and therapeutic radiation including exposure to radioisotopes	<10 rads 3.0% cases 3.4% controls 10–100 rads marrow dose 1.7% cases 1.3% controls 100+ rads marrow dose 2.3% cases 0.3% controls	Significant excess of therapeutic radiation for cases with acute leukemia (cell type not further specified), CML and CLL compared with controls.
Linos et al., 1980 [60]	138 leukemia cases constituting all cases diagnosed in Olmstead County, Minnesota (1955–74)	Diagnostic and low-level therapeutic radiation	< 300 rads	No association between low-dose radiation and myelocytic or lymphocytic leukemia. (A breakdown of each of these categories into acute and chronic did not alter the results.)
MacMahon 1962 [65]	304 children who died of leukemia (of 584 who died of all cancers) from an estimated 734,243 births, 1947–54, in 37 large maternity hospitals in the northeast U.S.	Diagnostic during pregnancy in mothers (radiation to the abdomen and pelvis)	Unknown	The ratio of observed to expected number of leukemia deaths (cell type not specified) was 1.32 (expected numbers computed from rates of x-ray exposure among a 1% systematic sample of the 734,243 births)

Table 6-9 Studies examining the relationship between therapeutic medical radiation and subsequent development of leukemia – follow-up of radiated (prospective)

Author(s), year	Study population	Reason for radiation	Site of radiation	Amt. of radiation	Findings with respect to leukemia
Diamond et al., 1973 [64]	10,169 white and 9720 black children exposed to intrauterine x rays in Baltimore, 1947–59	Diagnostic during pregnancy in mothers	X-ray pelvimetry during pregnancy in mothers	Unknown	Death rate from leukemia (cell type not further specified) was 3 times higher in exposed whites than controls. But there was no difference in the mortality experience of black exposed children compared with controls.
Modan, 1974 [72] Modan, 1977 [73]	10,902 children from North Africa and various mideastern countries who received scalp radiation for tinea capitis upon immigrating to Israel, 1946–60.	Scalp irradiation for tinea capitis	Scalp	Average was 9 rads	No leukemia excess. (Leukemia rate, all types combined per 1,000 was 0.7 for those irradiated compared with 0.5 in population controls and 0.4 in sibling controls.)
Albert and Omran, 1968 [66] Hurley et al., 1976 [67]	2043 children irradiated for tinea capitis at the Skin and Cancer Unit of New York Univ. Hosp. 1940–59	Scalp irradiation for tinea capitis	Scalp	6 rads to thyroid 49 rads to pituitary 39 rads to parotid	The incidence of all tumors was 10 times higher in the irradiated (0.69%) than in the controls (0.06%). Of the 14 malignancies occurring in the irradiated, 4 were leukemias (2 ALL, 1 AML, and 1 CML).
Hempelmann et al., 1967 [68] Hempelmann et al., 1975 [69]	2878 infants treated with x irradiation for thymic enlargement, before 1959.	Thymic irradiation	Thymus	Unknown	7 cases of leukemia (cell type not specified) were observed; 2.2 cases were expected. In sibling controls 2 cases were observed, 3.8 expected. Significance was not tested.

Saenger et al., 1968 [70]	33,888 patients from 26 medical centers in the U.S. Of these 18,379 were treated with I^{131}; 3311 were treated with I^{131} and surgery, 1946–64	Hyperthyroidism	Thyroid (concentrates I^{131})	Mean bone marrow dose was 7–15 rads	Age- and sex-adjusted rates for acute leukemia and chronic leukemia did not differ between those treated with I^{131} and those treated with surgery for hyperthyroidism. Patients treated with both surgery and I^{131} had a significantly higher incidence of leukemia than either of the other 2 groups. Observed leukemia mortality of those with hyperthyroidism is 50% higher than U.S. rates.
Pochin, 1969 [71]	215 patients with inoperable thyroid cancer treated with radioiodine, 1949–60	Treatment for inoperable thyroid carcinoma	Thyroid (concentrates radioiodine)	Unknown	Excess mortality of leukemia (cell types not specified)
Favus et al., 1976 [74]	1026 children (and a few young adults) who were treated with radiation for infectious and inflammatory disease of the upper respiratory tract, 1940–60	Treatment for infectious and inflammatory disease of the upper respiratory tract	Upper respiratory tract	Unknown	Leukemia incidence and/or mortality was not examined but there was a large prevalence of palpable, nodular thyroid disease (16.5%). Of those operated on in this group, 33% had thyroid carcinoma.
daSilva Horta et al., 1974 [83] daSilva Horta et al., 1978 [84]	1237 patients who received thorotrast (a contrast material) during cerebral or limb angiography procedures, 1930–55	Diagnostic evaluation	Injected intraarterially	Average amounts used were 20–39 ml (radiation dose not specified)	Excess incidence of acute leukemias (cell types not further specified) as well as aplastic anemia, malignant neoplasms of the liver, and liver fibrosis.

Table 6-9 (continued)

Author(s), year	Study population	Reason for radiation	Site of radiation	Amt. of radiation	Findings with respect to leukemia
Randall et al., 1964 [78]	6569 women treated for menorrhagia of which 3382 were treated with radiation, before 1940	Treatment for menorrhagia	Pelvis	Average amt. was 200 roentgens for x rays and 600–1800 mg-hours for radium	No excess of leukemia or mortality in general.
Doll and Smith, 1968 [75] Smith and Doll, 1976 [76]	2068 women treated for menorrhagia with radiation to induce an artificial menopause, 1940–60.	Treatment for menorrhagia	Pelvis	Average amt. was 500–1050 rads to each of 2 parallel opposed pelvic fields.	Excess mortality from leukemia (1.1 cases excess per million women per year per rad, cell types not specified separately) and cancer of heavily irradiated organs including ovaries, intestines, rectum, uterus, bladder.
Alderson and Jackson, 1971 [77]	2049 women in Manchester, U.K. treated with x rays or radium for menorrhagia, 1946–60.	Therapy for for menorrhagia	Pelvis	300–1500 rads depending on age and weight	Slight excess of leukemia (cell types not reported)
Kohn et al., 1965 [79] Zippin et al., 1971 [80]	497 women selected from Connecticut and California Cancer Registries who were treated for cancer of the cervix, 1931–51	Therapy for cancer of the cervix	Pelvis	X rays: 1–54 million gram-rads radium: $<$ 3.5 million gram-rads	No excess of leukemia. No statistically significant differences between study group and controls.

Table 6-9 (continued)

Hutchison, 1968 [81]	29,493 women from 29 clinics in 9 countries, prior to 1968	Therapy for cancer of the cervix	Pelvis	Average dose to bone marrow in the abdomen was 3000 rads	No significant excess of leukemia.
Sulzberger et al., 1952 [82]	1000 patients treated for dermatologic conditions with radiation at the Skin and Cancer Unit of NYU Hosp., 1933–47	Radiation therapy for dermatologic conditions	Skin	724/1000 received 150–1000 rads. 201/1000 received 1000–2000 rads. 75/1000 received >2000 rads	No excess of leukemia. 25% of cases had cancer and other malignant growths of skin.
Court-Brown and Doll, 1957 [86] Court-Brown and Doll, 1965 [87] Radford et al., 1977 [88]	13,352 patients with ankylosing spondylitis treated with radiation, 1935–54.	Radiation therapy for immobility of the spine	Spine	0–3000 rads to bone marrow	Excess leukemia (cell types not specified) as well as aplastic anemia, lymphoma, cancer of stomach, pancreas, lung, and bone. No excess of leukemia in persons with ankylosing spondylitis *not* treated with radiation.

for males in the Tri-State study and for both sexes in the Gunz and Atkinson study. Linos et al. found no association, but the number of cases (138, ages not specified) included in this investigation was very small and consisted of both adults and children representing all incident cases in Olmstead County, Minnesota, 1955–74 [58].

For children, an increased risk for childhood leukemia (cell type not specified) was associated with preconception x rays of both parents in the Tri-State study [53–58] and x-ray exposure to the fetus in utero received by pregnant mothers in the studies of Stewart et al. [61–63] and MacMahon [65]. Stewart et al. noted excess risk of ALL (R.R. = 1.6), AML (R.R. = 1.6), and other unspecified leukemia (R.R. = 1.3); leukemia cell type was not examined in the other two studies.

Prospective Studies In these studies, groups of individuals who received radiation for diagnostic or therapeutic reasons were followed up to determine their leukemia incidence and mortality. In Table 6-9, we have grouped some of these studies by anatomic site of exposure to radiation.

A study by Diamond et al. [64] found an excess of leukemia developing in children whose mothers had x-ray pelvimetry during pregnancy. These investigators found the death rate from leukemia (cell type not further specified) to be 3 times higher (although numbers of leukemia deaths were very small) in the exposed white children compared with controls but no higher in black children compared with controls.

Most of the prospective studies examining the risk of developing malignancies following radiation to the head and neck for a variety of conditions have found an excess incidence of leukemia as well as other tumors. Albert and Omran and Hurley et al. noted a tenfold excess for all tumors in a population of 2,043 children treated with scalp irradiation for tinea capitis in New York City. Among the 14 malignancies found by these investigators were four leukemias (2 ALL, 1 AML, and 1 CML) [66, 67]. In 2,878 children who received irradiation as infants for enlarged thymuses, Hempelmann et al. found 7 cases of leukemia (cell type not given) when 2.2 were expected [68, 69]. Excess mortality from leukemia was found in 3,311 patients with hyperthyroidism treated with both I^{131} and surgery [70] and 215 patients with inoperable thyroid cancer given radioactive iodine [71]. Of interest was the finding by Saenger et al. that leukemia mortality in those having hyperthyroidism treated with either I^{131} alone or surgery alone showed no difference in age-adjusted acute or chronic myelocytic leukemia rates. Comparison of combined leukemia rates in persons with hyperthyroidism who developed leukemia both *prior to* and following all three types of treatment (i.e., I^{131} alone, surgery alone, or both I^{131} and surgery) with age- and sex-adjusted U.S. population rates showed a 1.5-fold excess in the hyperthyroid population [70]. Although it is possible that this excess could reflect better surveillance or use of more sophisticated diagnostic tests in this patient population, it is also feasible that the increased mortality might be related to the underlying condition of hyperthyroidism. In contrast to these studies demonstrating an association, Modan et al. found no leukemia excess in a follow-up study of 10,902 children from North Africa and various Middle Eastern countries who, upon immigrating to Israel, received scalp irradiation for tinea capitis [72, 73]. Another apparently negative study was described by Favus et al. who reported no excess of leukemia in 1,026 children and young adults who had been treated with x irradiation for infectious and/or inflammatory disease of the upper respiratory tract [74].

A number of investigators have studied populations of women treated with radiation for menorrhagia to halt the heavy bleeding. Two studies did find an excess rate of leukemia [75, 76]. Doll and Smith calculated that the excess was 1.1 per million women per year per rad. These investigators also noted an increased incidence of cancer developing in heavily irradiated, adjacent organs including ovaries, uterus, intestines, rectum, and bladder [75, 76]. Randall et al. found no excess of leukemia mortality in 3,382 women receiving radiation treatment for menorrhagia [78].

Studies of women treated with radiation for cancer of the cervix found no excess of leukemia. The studies of Kohn et al. and Zippin et al. of 497 women identified from the Connecticut and California Cancer Registries showed no significant leukemia mortality differences between cervical cancer cases and controls although the number of cases was rather small [79, 80]. A large follow-up study from 29 clinics in nine countries was conducted by Hutchinson on 29,493 women treated with radiation for cervical cancer. He also found no significant excess of leukemias even though the average radiation dose to bone marrow in the abdomen was 3,000 rads [81].

Sulzberger et al. found no excess of leukemia in follow-up of 1,000 patients treated with radiation therapy for dermatologic conditions [82].

Other types of radioactive substances used for both diagnostic and therapeutic purposes have also been found to be associated with an excess risk of acute leukemia. These include thorotrast, a radioactive contrast material used many years ago, which was also noted to result in increased incidence of aplastic anemia, primary malignant neoplasms of the liver, and liver fibrosis [83, 84]. Also, Modan and Lilienfeld showed an increased risk of acute leukemia developing in patients with polycythemia vera treated with P^{32} compared to those treated with phlebotomy [85].

One of the earliest and most important prospective studies of population groups treated with radiation for therapeutic purposes was the 13,352 men with ankylosing spondylitis treated during 1935–54. Court-Brown and Doll found an excess of leukemia (cell types not specified) as well as aplastic anemia, lymphoma, carcinoma of the pancreas, stomach, and bone [86, 87]. However, persons with ankylosing spondylitis not treated with radiation do not appear to have an increased risk of leukemia [88].

Summary The risk of leukemia from high doses of radiation, such as those to which Japanese atomic-bomb survivors close to the bomb were exposed (50+ rads), is well established. Radiation exposure in this population was associated with increased risk of ALL, ANLL, and CML but not CLL. While cell types of excess leukemia are not specified in many studies, the findings of Caldwell et al. (ALL, AML, CML), Mancuso et al. (myeloid leukemia), the Tri-State Study (AML, CML), and Albert et al. (ALL, AML, CML) support the findings from the Japanese atomic-bomb survivor studies. All cell types except CLL appear to be associated with radiation. Groups receiving fairly high radiation exposures as a result of medical therapeutic exposures also frequently appeared to have an increased risk. Examples of such groups include the patients with ankylosing spondylitis treated with radiation to the spine and populations treated with radiation for menorrhagia in order to halt bleeding or introduce an artificial menopause. However, women treated with substantial amounts of radiation for cancer of the cervix did not have excess leukemia incidence or mortality.

The association of leukemia and low-dose prenatal x ray exposures is fairly well established in whites; there does not appear to be an excess in exposed black children. Associations between treatment with radiation for thymic enlargement, some groups with tinea capitis, and thyroid cancer have also been noted. Use of thorotrast as a contrast material has also been shown to result in increased leukemia incidence. Excess leukemia has also been noted in some populations of radiologists (especially those heavily exposed during the first half of the twentieth century) and in children in Utah exposed to fallout from nuclear test explosions carried out in the 1950's.

While radiation is one of the best-studied carcinogens, additional studies of large populations exposed to low levels could be helpful in further defining the shape of the dose-response curve at low-exposure levels.

OCCUPATIONAL EXPOSURES, CHEMICALS, AND THE LEUKEMIAS

Evidence has accumulated during the past 20 to 30 years that the environment, including occupational exposures, may play a substantial role in cancer causation. Higginson compared high and low incidence rate regions of the world for various cancers and suggested that substantial human cancer incidence can be ascribed to environmental factors [89].

Wynder and Gori examined the ratio of high to low risk incidence and mortality rates for different malignancies and found it to be greater than 2 for all except leukemia. For males, the age-adjusted leukemia mortality ratio of Denmark (high risk) compared to Japan (low risk) was 2.1:1; for females, the ratio of Israel (high risk) compared to Japan (low risk) was 1.9:1. Because environmental factors do not affect all population groups equally, Wynder and Gori assumed that populations and population subgroups with the lowest rates would have the least exposure to specific etiologic factors [90]. Since incidence rates between and within countries, and between sexes can differ substantially, these investigators believe that epidemiologic evidence lends support to Higginson's conclusion.

Occupational exposures are an important aspect of the environment. Data are insufficient to estimate the proportion of each of the leukemias that are due to occupational exposures. However, Enterline suggests that data from studies of occupational exposures may be generalizable to other similar enviromental but nonoccupationally related exposures [91]. Occupational data can be used to estimate the effect of continuous environmental, nonoccupational exposures in the general population by converting 8-hour, 5-day week occupational exposures to 24-hour, 7-day week exposures. As an example, the Carcinogen Assessment Group, a federal government interagency group of scientists and statisticians, used data from occupational epidemiology studies and a linear, nonthreshold model to estimate the leukemia risk from public and community exposure to benzene, exclusive of workplace exposure [92].

Benzene

First in 1928 and later in 1938, cases of leukemia developing subsequent to benzene exposure were reported [93, 94]. In 1939, Mallory described leukemia in autopsies of individuals exposed to benzene [95]. Browning reported 61 cases of leukemia believed to be related to benzene in 1965 [96]. By 1976, Vigliani and Forni had collected over 150 reports of benzene-associated leukemia cases. These authors described, by histopathologic type, 86 of these 150 cases; 60 were acute leukemia (not further specified as to ALL or ANLL), 3 were erythroleukemia, 8 were CLL, and 13 were CML [97]. Vigliani described the occupations of 11 leukemics among 66 cases of "chronic benzene poisoning," seen at the Institute of Occupational Medicine of Milan. The majority were employed in the shoe manufacturing, rotoprinting, rubber, and leather industries [98]. Brandt et al. noted that three cases of ANLL who had a negative medical history of benzene exposure were found to have had substantial exposure upon careful re-evaluation of events preceding the development of leukemia [99].

Case reports are difficult to interpret but epidemiologic studies support the association between benzene and leukemia. Only some of these studies examining associations between leukemia and benzene have been histopathologic-type specific. Brandt et al. carried out a case-control study comparing 50 cases of males between ages 20 and 65 with ANLL to 100 male controls with the same age distribution who were treated at the same internal medicine clinic for allergic conditions. Thirty-six percent of those with ANLL reported occupational exposure to petroleum products (of which benzene is an important constituent) compared to 10% of controls. Data from the Swedish census of 1970 indicates that approximately 10% of the working male population in that age range reported occupational exposure to petroleum, similar to the controls used in the study of Brandt et al. No difference in occupational exposure to petroleum was reported between CLL and CML cases and controls [100].

These investigators also compared histopathologic type and chromosome analyses of 23 ANLL cases occupationally exposed to chemical solvents, insecticides, and petrol products to leukemic controls without similar occupational exposures. AML was a more common histopathologic type in the exposed group (74%) than the unexposed (42%). Neither the remission rate nor the median survival differed significantly between the two groups. Of the exposed leukemics, 82.6% had chromosomal aberrations compared with 24.2% of the nonexposed group. Four main chromosomal abnormalities distinguishing the exposed group were monosomy 5, monosomy 7, trisomy 8, and trisomy 21. Two of these were not found in controls; the two others were each noted in one patient in the control group [101].

In Istanbul, Turkey, Aksoy et al. identified all cases of leukemia diagnosed 1967–73 at the hospitals affiliated with two medical schools. Twenty-six cases of leukemia had developed in persons employed in the shoe leather industry. The population at risk was estimated to be 28,500 shoe leather workers. The leukemia incidence represented a 20-fold excess compared with the age-adjusted leukemia rate in the general population. Latent periods were estimated to range from 1 to 15 years. Although the estimate of the population at risk may have been imprecise and follow-up may have been incomplete since other hospitals, physician records, and death certificates were not studied, other studies support Aksoy's findings [102]. Aksoy et al. also examined the distribution of the various types of leukemia due to benzene exposure for the 26 cases in his follow-up study as well as eight additional leukemics with a history of benzene poisoning. Eight-five percent had ANLL, 12% had ALL, and 3% had CML [103].

Infante et al. followed up the vital status of 748 workers until 1975 who had been employed 1940–49 at two pliofilm manufacturing plants and discovered seven cases of ANLL (later increased to nine cases). The nine cases represented a sevenfold excess [104, 105]. Neither this study nor that of Aksoy measured average exposure to workers.

Ott et al. carried out a follow-up study of vital and occupational status through January 1, 1974 of 594 workers employed during 1940–73 in the Michigan Division of the Dow Chemical Company. The three cases of ANLL found represented a 3.5-fold excess compared to U.S. population age-adjusted ANLL rates. Average benzene exposure in this population was measured at 9–11 ppm with peak exposures of 250–937 ppm [106].

Thorpe found no excess leukemia in a follow-up study during 1962–71 of 38,000 workers employed in a variety of petroleum and petrochemical operations in eight affiliates of the Exxon Corporation. The "exposed" group was defined as individuals who worked for five or more years in jobs where there was potential exposure to 1% or more benzene. No leukemia excess was noted when separately analyzed by histopathologic type. A number of methodologic problems characterized this study. The population at risk was limited to active workers and those on pensions; persons who left the company prior to pension eligibility were not included. In addition, there was inadequate follow-up of those on pensions. The available occupational histories for leukemia cases were incomplete. Cases diagnosed with leukemia were not validated nor were those with other diagnoses who had possibly been misclassified [107]. Another large follow-up study of petroleum refinery workers also showed no excess of leukemia [108].

A case-control study by Ishimaru et al. examined the association between benzene and leukemia for subgroups with various levels of radiation exposure. These investigators compared 303 cases of adult leukemia with onset 1945–67 in Hiroshima or Nagasaki selected from the Atomic Bomb Casualty Commission (ABCC) Leukemia Registry with controls selected from ABCC-fixed samples (matched by city, sex, date of birth, exposure distance, vital status, and residence in either Hiroshima or Nagasaki City at time of onset of disease) for history of benzene and medical x-ray exposure. Cases were classified into three radiation exposure categories. For the high radiation exposure group ($\geq$100 rads), the risk of leukemia was 1.5 times greater among those with a history of exposure to benzene and/or medical x rays compared to matched controls. The low radiation group (0–99 rads) had an estimated relative risk of 4.0; the unirradiated group had an estimated relative risk of 2.4 [109].

Other benzene-containing products have also been associated with leukemia. Several reports noted development of leukemia following exposure to chlorinated insecticides, many of which contain a significant amount of benzene [110–113]. CLL has been reported to occur in several members of a family, all of whom were employed in a dry-cleaning business [114, 115]. Since benzene is such a ubiquitous chemical from which many products are made, it is possible that a substantial proportion of leukemias may be related to this substance.

McMichael et al. found an association among rubber workers between CLL and exposure to such solvents as benzene, toluene, and a variety of other organic agents. A cohort mortality study showed a sevenfold excess of deaths from CLL at Akron, Ohio, rubber plants compared to a U.S. standard population and a small mortality excess at other non-Akron rubber plants. A case-control study was carried out, using those who had died of CLL as cases, to attempt to determine relationships of CLL to occupational title and whether leukemia and related cancers are associated with exposure to solvents. A statistically significant positive association was found between solvent exposure and CLL for three solvent-exposure occupational title groups (one with heavy, a second with moderate, and a third with light solvent exposure). The relative risk of developing CLL following heavy solvent exposure was 5.5, medium solvent exposure, 2.0, and light exposure, 2.5. There was a longer exposure period in the solvent-exposure jobs for cases than for controls, 11.27 in contrast to 4.06 years. Close examination of work histories in those dying of CLL showed that the greatest difference between cases and controls was for jobs in the tire repair occupational title; the relevant exposure agent(s) for workers in that job category included benzene, xylene, toluene, trichlorethylene, and various aliphatic hydrocarbons [116–118].

Monson et al. found an SMR of 160 for lymphoproliferative malignancies and multiple myeloma and an SMR of 150 for leukemia among rubber company employees who worked in the tire-building area of the company. Lymphoproliferative malignancies and leukemia were not further classified into cell types [119].

Fabia and Thuy reported that fathers of children with leukemia and lymphoma (grouped together) were more likely to be employed in hydrocarbon-related occupations (motor-vehicle mechanic, service station attendant, odds ratio [O.R.] = 2.2; machinists O.R. = 2.2; painters O.R. = 2.2) than were fathers of controls [120]. This possible association was further examined by Hakulinen et al. in a case-control study of 852 children identified through the Finnish Cancer Registry and 852 matched controls. No significant association was reported between leukemias and lymphomas as a group and employment of fathers in hydrocarbon-related industries (children under age 5 at age of diagnosis, O.R. = 0.68; children under 15 at diagnosis, O.R. = 0.95) [121]. Zack et al. also found no difference in frequency of parents employed in hydrocarbon-related industries in childhood cancer cases compared to controls. The cancer cases were not further analyzed by specific cancer cell types [122].

Exposures other than benzene may be involved in the apparent excess of leukemias and lymphomas occurring in 530 men who graduated from the School of Chemical Engineering at the Royal Institute of Technology, Stockholm [123]. The mechanisms by which solvent exposures (particularly benzene and other polycyclic hydrocarbons) can cause leukemia are not well understood. Sugiyama has found reproducible chromosomal changes in animals with polycyclic hydrocarbon-induced leukemia. Analysis of 345 cases of leukemia induced in outbred Long-Evans and Sprague-Dawley rats by polycyclic hydrocarbons (7,12-dimethylbenz(*a*)anthracene and 7,8,12-trimethylbenz(*a*)anthracene) has shown 24% of the cases to have stemline leukemia cells with a #2 trisomy, 3% had leukemia cells with a long #2 chromosome, 10% had polyploid leukemias or other markers, and 63% had a normal diploid karyotype. These investigators found that the incidence of the #2 abnormalities depended on the chemicals used for leukemia induction, which suggested that the structure of the chemical carcinogens influences the formation of the chromosome changes [124].

Chromosome aberrations have been reported in humans exposed to solvents such as benzene, toluene, and vinyl chloride; many of these reports are limited to exposed case groups without any control comparisons [125–128]. A study by Forni et al. found a significantly higher proportion of stable and unstable chromosome aberrations in a benzene-exposed group of workers compared with controls. These investigators also noted a significant difference in the frequency of chromosome abnormalities in the benzene-exposed group compared to a toluene-exposed group and another control group not exposed to solvents [129].

In an unpublished study, scientists at Dow Chemical Company found significant chromosome abnormalities in workers exposed to 3–9 ppm benzene over a number of years as compared to controls but there was no evidence of overt disease in the exposed group. How frequently significant chromosome abnormalities are transformed into one or more diseases is not known. A need for long-term follow-up of these workers and those studied by Forni et al. is indicated. It is particularly important in the case of the Dow workers because of the very low exposure levels of these workers to benzene [130].

Asbestos

Lieben noted 21 malignancies recorded in medical charts of 68 (31%) workers from an asbestos-producing plant who were diagnosed as having asbestosis. Five of the 21 had lymphoproliferative malignancies; of these five, two had chronic lymphocytic leukemia. Controls were not used in this study [131]. Gerber described a series of 35 cases of asbestosis proved by autopsy. Five individuals in this series had tumors of the hematopoietic system. The method by which cases in the series were selected was not specified and controls were not used [132]. Additional studies, especially those using epidemiologic methods, are required to further delineate the relationship between asbestos exposure and the leukemias.

Farming

Farmers have been repeatedly shown to have an excess incidence of leukemia, particularly lymphatic leukemia, but also acute leukemia [133–135]. The exposures that increase the farmers' risk have not been completely identified. Blair et al. found an increased risk of acute unspecified leukemia among farmers in heavy corn, hog, chicken-producing, and herbicide-using counties in Nebraska. However, the risk was not high in counties with heavy pesticide use in the absence of the other factors, such as high corn production [135].

Electrical and Magnetic Fields

Milham has noted increased proportionate mortality ratios for leukemia in men whose occupations required work in electrical or magnetic work fields. Occupational groups with more deaths due to leukemia than expected included electricians (all leukemia PMR = 138; acute leukemia, not otherwise specified, PMR = 178), television and radio repairmen (all leukemia PMR = 157; acute leukemia PMR = 291), power station operators (all leukemia PMR = 259; acute leukemia PMR = 282), aluminum workers (all leukemia PMR = 189; acute leukemia PMR = 258), and a number of other similarly exposed occupational groups [136]. Three other studies have found elevated proportionate mortality or proportionate incidence ratios for acute myelocytic leukemia [137, 138] or all leukemias combined [139] in a variety of electrical occupations. However it is unclear, given the focus on PMR or PIR as a measure, what these findings mean [140]. This association is also supported by a study of Wertheimer et al., who compared 155 cases of childhood leukemia (histopathologic type not further specified) to controls and noted that 38.2% of cases had a birth address in close conjunction with high-current electrical wiring, compared with 21.3% of controls. Similarly, 40.6% of cases had a death address in conjunctions with high-current wiring, compared to 18.7% of controls [141]. But Fulton et al. found no association between childhood leukemia and electric power line configurations [142].

A summary of the occupations and associated exposures studied epidemiologically and suspected of having an association with leukemia is shown in Table 6-10.

A case-control study comparing occupations of 1,845 men with leukemia identified in the Tri-State Leukemia Study with those of 1,237 control males from a random sample of the same geographic area suggested a somewhat higher risk of leukemia in the construction industry. Further analysis showed that painters were primarily responsible for the excess risk with a risk ratio of 282 for all ages; for those 15–44 years, it was 2.20;

for those 45–64 years, RR = 3.29; and for those 65–98 years, RR = 2.90. No relationship was noted between leukemia and occupation among women either for industries or by single occupation. Separate analysis by leukemia subtype was not reported [143]. Additional case-control studies, particularly those examining a single histopathologic-specific leukemia subtype, would be useful. Since occupations of women have changed extensively in the past 10–20 years, separate analysis by sex would be very important.

A number of analyses of occupational influences on disease mortality and incidence have used occupational data from death certificates or occupational information obtained from interview of a sample of cancer cases included in the Third National Cancer Survey. These analyses consist of cross-sectional data and hence should be interpreted cautiously. However, such studies may suggest clues for hypothesis testing in more focused case-control and prospective cohort studies.

Occupational mortality data from the Registrar General's Decennial Supplement, England and Wales, 1971, showed several occupations with significantly elevated SMRs for leukemia (not further classified) or for neoplasms of lymphatic and hematopoietic tissue (not further specified). These are shown in Table 6-11 [144].

De Coufle et al. conducted a retrospective study of the lifetime occupation of specific groups of cancer cases, including leukemia, seen at the Roswell Park Memorial Institute in Buffalo, New York, 1956–65, compared with relative frequency of occupations of a control group consisting of individuals referred to Roswell Park during the same years diagnosed with non-neoplastic diseases. However, leukemia was not separately analyzed by histopathologic type. Results presented in Table 6-12 show that risks for managers and farmers which were significantly increased in data reported by the Registrar General are not excessive in this analysis [145].

Occupational and cause-of-death information on approximately 300,000 death certificates in Washington State among males in 1950–71 was examined by Milham to ascertain the mortality pattern for different occupational groups. Causes of death were coded mainly according to the 7th Revision. Since population data were not available by occupations, Milham calculated proportional mortality ratios. Occupations characterized by elevated PMRs for leukemia are shown in Table 6-13 [146]. PMRs are elevated for a number of different types of farmers; lawyers and judges; and professional workers of different types as were the SMRs for these groups in the Registrar General's analysis. However, PMRs may not always reflect mortality patterns derived using population-based statistics.

In the U.S. Third National Cancer Survey, an interview study of 7,518 incident cancer cases were carried out. Lifetime histories of occupations

Table 6-10 Summary of studies showing an association between leukemia and occupation

Author, year	Occupation	Exposure	Leukemia type	Risk ratio	Comment
Brandt et al., 1978 [100]	Not stated	Petroleum products	ANLL CLL CML	3.6 No excess No excess	Case-control study with cases of leukemia from one clinic.
Aksoy et al., 1974 [102]	Shoe-leather workers	Benzene	ANLL (85%) ALL (12%) CML (3%)	20	All cases were identified at hospitals of 2 medical schools. Population at risk was estimated.
Infante et al., 1977 [104]	Pliofilm workers	Benzene	ANLL	7	Follow-up study.
Ott et al., 1978 [106]	Chemical company workers	Benzene and other carcinogens	ANLL	3.5	Follow-up study.
Thorpe, 1974 [107]	Petrochemical company workers	Benzene and other solvents	Leukemia as a group and specific histopathologic types	No excess	Follow-up study: retired pension and nonpension workers incompletely followed up.
Ishimaru et al., 1971 [109]	Not stated	Benzene and medical x rays	Leukemia as a group	2.5	Case-control study of leukemics from the ABCC leukemia registry, Hiroshima and Nagasaki
McMichael et al., 1974, 1975, 1976 [116–118]	Rubber workers	Solvents	CLL	3.5	Follow-up study.
Monson et al., 1976 [119]	Rubber workers	Employes who worked in the tire-building area—exposures not further identified	Leukemia as a group	SMR 150	Follow-up study.

Fabia et al., 1974 [120]	Fathers of children employed in hydrocarbon-related industries	Hydrocarbons	Leukemia and lymphoma		Case-control study of children with malignant disease dying before age 5 years.
	Motor-vehicle mechanic and service station attendant			2.2	
	Machinist			2.2	
	Painter			2.2	
Hakulinen et al., 1976 [121]	Fathers worked in hydrocarbon-related industries	Hydrocarbons	Leukemia and lymphoma	No excess	Case-control study listing 852 cancer cases from Finnish Tumor Registry.
Zack et al., 1980 [122]	Patients worked in hydrocarbon–related industries	Hydrocarbons	All cancers; not analyzed separately by type	No excess	Case-control study with 296 children from hematology clinic.
Olin, 1976 [123]	Graduates of School of Chemical Engineering	Not stated	Leukemias and lymphomas	3.5	Follow-up study.
Fasal et al., 1968 [133]	Farmers	Unknown	All leukemias	SMR = 114 for both males and females	Study comparing causes of death for farm and non-farm residents using death certificates.
Milham, 1971 [134]	Farmers	Unknown	Lymphatic (7th Revision) Acute (7th Revision) Acute unspecified (7th Revision)	O.R. not given	Excess of lymphatic leukemia and acute leukemia deaths among farmers under age 60.
Blair et al., 1979 [135]	Farmers	Unknown; counties with high inventories or production of: cattle, corn, hogs or herbicides	Acute unspecified (7th Revision)	2.21	Increased risk limited to farmers dying before age 66.
			Lymphatic leukemia (7th Revision)	2.00	
			Acute unspecified:		
			Cattle	2.27	
			Corn	1.75	
			Hogs	1.90	
			Herbicides	2.00	

Table 6-11 Occupations with high SMRs for leukemia or neoplasms of lymphatic and hematopoietic tissue – males

Occupation	Neoplasm	SMR	p value
Architects, town planners	Leukemia	278	< 0.01
Managers (not elsewhere classified)	Leukemia	136	< 0.05
Ministers of the Crown, MPs, senior government officials	Leukemia	218	< 0.05
Farmers, farm managers, market gardeners	Lymphatic and hematopoietic neoplasms	131	< 0.05
Judges, barristers, solicitors	Lymphatic and hematopoietic neoplasms	217	< 0.05
Professional workers	Lymphatic and hematopoietic neoplasms	162	< 0.05

Source: Registrar General's Decennial Supplement [144].

and industries were obtained and examined for associations with specific cancer sites and types with adjustment for age, sex, race, education, consumption of alcohol, smoking, and geographic residence. Both occupation and industry were coded for each subject according to the 1970 census coding scheme with separate entries for main lifetime employment, recent employment, and other jobs held. The main analytic approach consisted of serial comparison of the proportions of specific main lifetime industries and occupations among patients with cancer at one site with those of patients having cancers at other sites combined as a control group. Results for leukemia are displayed in Table 6-14 [147].

The cell-specific data are presented only for industries rather than for particular occupations, thereby making it difficult to compare with data in Tables 6-12 and 6-13. Nevertheless, elevated odds ratios for leukemia were

Table 6-12 Occupations with high or low risk ratios for leukemia – males

Occupation	Risk ratio	p value
Machinists	2.847	0.02
Operatives in primary metal industries	2.159	NS
Mine operatives and laborers	2.164	NS
Farmers	0.585	NS
Carpenters	0.780	NS
Foremen, all industries	0.793	NS
Operatives in miscellaneous manufacturing industries	0.600	NS
Mechanics and repairmen	0.751	NS
Truck and tractor drivers	0.566	NS
Bus, taxi, and truck drivers	0.707	NS
Farmers and managers in general crops and livestock	0.546	NS

Source: DeCoufle et al. [145].

Table 6-13 Occupations with high PMRs for leukemia — males. Washington State occupational and cause of death information from death certificates

Occupation	Type of leukemia	PMR
Accountants	Acute leukemia	220
Airplane pilots	All leukemias	203
Chemists	All leukemias	243
Funeral directors and embalmers	All leukemias	299
Lawyers and judges	Myeloid leukemia	277
School teachers	Myeloid leukemia	240
Students	Myeloid leukemia	308
Farmers	Lymphatic leukemia	135
Ranchers	Myeloid and acute leukemia	130
Wheat and grain farmers	All leukemias	147
Poultry farmers	All leukemias	269
Store buyers	All leukemias	341
Officials: lodge, society, union	Myeloid leukemia	159
Bankers	All leukemias	183
Sales clerks	Acute leukemia	188
Electricians	Acute leukemia	184
Mechanics and repairmen	Myeloid leukemia	173
Millwrights	Acute leukemia	204
Stationary engineers	Myeloid leukemia	247
Aircraft workers employed by Boeing	Acute leukemia	168
Creamery workers, dairymen	Myeloid leukemia	330
Packers and wrappers	All leukemias	379
Plywood workers	Myeloid and acute leukemia	275
Firemen	Lymphatic leukemia	233

Source: Milham [146].

Table 6-14 Industries with high risk ratios for leukemia by cell type — males and females. Third National Cancer Survey occupational history interview

Industries	Sex	Type of leukemia	Odds ratio
Construction	M	CLL	2.37
Manufacturing	M	ALL	1.92
	M	CLL	2.17
Utilities and communications	M	CLL	2.45
Sales	M	ALL	5.19
	M	AML	2.94
	M	CML	7.39
	F	AML	4.45
Finance, insurance and real estate	M	CML	7.76
Other personal services	M	AML	12.84
	F	CLL	4.38
	F	AML	1.28
Professional and related services	F	CLL	1.62
Public administration	M	AML	1.33
Medical services	F	CLL	2.13
	F	AML	2.35

Source: Williams et al. [147].

Table 6-15 Summary[a] of occupations and industries associated with the leukemias by histopathologic type

Occupations and industries	Leukemia type	Comment
Manufacturing	ALL	Males only
Sales		Males only
Shoe leather workers	ANLL	Males only
Pliofilm workers		Males only
Chemical company workers		Males only
Sales		Males and females
Other personal services		Males and females
Public administration		Males only
Medical services		Females only
Sales	CML	Males only
Finance, insurance, and real estate		Males only
Rubber workers	CLL	Males only
Construction		Males only
Manufacturing		Males only
Utilities and communications		Males only
Other personal services		Females only
Professional and related services		Females only
Medical services		Females only

[a]Based on Tables 6-10, 6-13, and 6-14.

observed for persons employed in sales, finance, and various professional services, a finding consistent with the elevated PMRs for these occupational groups in the Washington State data.

Although data from cross-sectional analyses may not be as reliable as data from case-control and particularly prospective studies, especially for associations between occupation and disease, it may be useful to summarize all of the occupations and industries found to be associated with the leukemias separately by histopathologic type. These are shown in Table 6-15; measures of association have been omitted because they are not comparable between studies. It is notable that farming and other occupations shown to be at increased risk for leukemia are not included in the summary Table 6-15 because studies of these groups have not coded leukemia deaths or cases in farmers and some of the other occupational groups by histopathologic type.

DRUGS

A number of drugs have been reported to be associated with one or more of the leukemias; the majority of these are case reports. However, a few systematic studies have examined associations of leukemias and specific drugs. Most of the larger studies have investigated the role of cancer chemotherapy and immunosuppressive drugs in the development of leukemia in persons with malignant and nonmalignant disorders; these are discussed in Chapter 9. Otherwise, there is a paucity of analytic studies. Other drugs that have been implicated in the onset of one or more cases of leukemia are summarized in Table 6-16. Evidence for two of these drugs, phenylbutazone and chloramphenicol, is discussed in detail.

Methodologic Considerations

When designing and analyzing epidemiologic studies of drugs potentially associated with leukemias, the methodological problems are similar to those relevant to epidemiologic studies of any suspected environmental factors. There may be long latent periods between initial use of a drug and the subsequent development of leukemia. One must be concerned with small effects; even a very low incidence of leukemia is important and sample sizes must be appropriate. Other carcinogens may be administered simultaneously with a particular drug, such as diagnostic or therapeutic radiation, special diets, etc. On the other hand, certain issues may be unique to considering drugs as a risk factor; for example, drugs are usually given to ill people and various illnessess may make the individual much more sensitive to the drug and promote the development of leukemia.

Experimental studies of leukemogenesis in animals due to ingestion or instillation of drugs need to be critically evaluated. Was an appropriate choice of animal species or strain made? Was sample size adequate? Were controls used and, if so, were they appropriately selected? Was histopathologic evaluation carefully performed and interpretations confirmed by independent blinded review? Were animal-to-human extrapolations justified on the basis of the evidence? Were appropriate adjustments made with respect to the latent period? What is the role of age, sex, and strain of animal used in leukemogenesis? Were genetic and host factors considered? What was the role of diet and other health factors? If a particular drug results in leukemia in an animal, it does not mean that a human will similarly respond but it does imply that the substance is suspect and therefore potentially leukemogenic for man. One must not forget that negative results in animals do not necessarily exclude the possibility that the substance is leukemogenic for man [148–150].

In 1957, Juhasz et al. reported the occurrence of six leukemias and one reticulum cell sarcoma in 50 white mice (inbred strains but not further described by the investigators) who received 82 mg total dose of Isoniazid (INH), an important drug used in the treatment of tuberculosis. There

Table 6-16 Drugs and chemical agents associated with leukemia (excluding cancer chemotherapy and immunosuppressive agents)

Author, year	Drug	Type of leukemia	No. of cases	Dose and duration
Berry, 1966 [154]	Amphetamine (Benzedrine[R])	AML	1	8–16 tablets per day for 2½ years
Welch et al., 1954 [155] Davies et al., 1967 [156] Cohen et al., 1967 [157] Fraumeni Jr., 1967 [158] Brauer et al., 1967 [159] Humphries, 1968 [160] Goh, 1971 [161] Cohen et al., 1973 [162]	Chloramphenicol (Chloromycetin[R])	AML CML Leukemia, not further specified	21 2 11	Total dose: 3.5–175 g
Cassileth, 1967 [163]	Chlorpromazine	AML	1	25–200 mg daily for 9 days
König et al., 1967 [164]	Griseofulvin (Fulvicin[R])	CML	1	500 mg daily for 2 months
Grossbard et al., 1968 [165] Garson et al., 1969 [166]	Lysergiside (LSD)	ANLL	2	50–300 mcgm dose for 5–15 doses

Sjoberg et al., 1965 [167] Perers et al., 1966 [168]	Oxyphenbutazone (Tandearil[R])	Acute leukemia	2	1–1.9 g
Bean, 1960 [169] Cadman, 1962 [170] Woodliff et al., 1964 [171] Chalmers et al., 1964 [172] McCarthy et al., 1964 [173] Dougan et al., 1965 [174] Leavesley et al., 1969 [175]	Phenylbutazone (Butazolidin[R])	AML ALL CLL CML Acute leukemia, not otherwise specified	24 6 1 2 5	100–600 mg per day for 6 days–4 years
Aksoy et al., 1974 [176]	Propylthiouracil	AML	1	150 mg per day for 2 years
Brod, 1959 [177] Schneider et al., 1970 [178]	Tolbutamide (Orinase[R])	AMoL	2	0.25–1.5 g per day for 8–22 months
Frisch, 1973 [179]	Trimethoprim-sulfamethoxazole (Septra[R])	?	1	?

Source: Swanson and Cook [180].

were no malignancies in the 50 white mice controls [151]. Other investigators also using large doses of INH have produced leukemia in R3 mice [152] and BALB/c inbred mice [153]. Similar associations have not been found in humans although methodologically sound epidemiologic studies have not been carried out. It must be emphasized that leukemias occurring in animals appear to differ in many respects from those arising in man and therefore extrapolation may not be altogether appropriate.

From Table 6-16, it can be seen that most of the drugs have been associated with one or two cases of leukemia. Except for phenylbutazone and chloramphenicol, most of the drugs possibly associated with leukemia have not been studied epidemiologically, and these two have been analytically examined in only a few epidemiologic studies. Nevertheless, a summary of these reports is presented in the hope that it may encourage additional study [154–180].

Phenylbutazone

A number of case reports linking use of phenylbutazone, an anti-inflammatory medication usually prescribed for joint pain, with subsequent development of leukemia (particularly ANLL) began to appear around 1960 [169, 172, 173, 181, 182]. Dougan and Woodliff summarized 27 cases in 1965. Most of these persons had initially taken phenylbutazone less than four years prior to the diagnosis of leukemia [174]. Ten of the 27 leukemia cases had ANLL. In addition to these cases reports of leukemia, a number of case-control studies have shown that phenylbutazone can depress bone marrow function as well as cause chromosomal aberrations [155, 183–185].

Compared with the large number of analytic studies examining the association between various other primary malignancies, their treatment, and subsequent occurrence of ANLL as discussed in Chapter 9, only two analytic studies have examined the association between phenylbutazone use and the subsequent development of ANLL. Friedman and Ury followed up through 1976 140,000 Kaiser-Permanente Medical Plan subscribers who received Kaiser pharmacy prescriptions for phenylbutazone during 1969–1973. Phenylbutazone was prescribed for 3,660 persons. Eight of these persons developed leukemia or lymphoma, whereas 9.7 cases were expected; one case of acute myelocytic leukemia occurred when 1.6 was expected [186]. Using the same population base, Friedman also carried out a case-control study of 127 cases of acute myelocytic leukemia in persons over 30 years of age compared with two sets of controls: matched hospitalized controls and matched Kaiser-Permanente Plan members [187]. While prior use of phenylbutazone was higher in cases compared with the

Plan member controls, but not with the hospital controls, the difference was not statistically significant. Friedman felt that the association between phenylbutazone and AML could be explained by an increased frequency of prior musculoskeletal disease in cases compared with member controls.

As Friedman pointed out, a true association may exist in spite of the negative findings. The association may have been obscured by changing patterns of prescription by Kaiser physicians who may have prescribed lower and fewer doses of phenylbutazone following publication of the case reports linking it to ANLL. If phenylbutazone were a leukomogen without a threshold level at which leukemogenesis occurs, it is possible that larger sample sizes than employed by Friedman and co-workers in the 1970's would be required to detect an association. Although lower and fewer doses of phenylbutazone are now prescribed than in the 1950's and 1960's when the first case reports were published, phenylbutazone is still prescribed fairly frequently, for a variety of musculoskeletal disorders, and therefore further study of a possible association is warranted.

Chloramphenicol

Fraumeni carried out a follow-up survey of 151 cases reported to have bone marrow depression following use of chloramphenicol [127], phenylbutazone [24], or both drugs. These 151 cases represented 66% of cases reported by physicians to the Registry on Adverse Reactions of the American Medical Association and the Adverse Reactions Branch of the Food and Drug Administration, 1954–65. (The other 34% of cases did not participate in Fraumeni's survey.) Leukemia subsequently developed in three patients whose bone marrow depression was associated with the use of chloramphenicol. The latent periods for the three cases, 1 AML, 1 CML, and 1 ALL, were 2½ years, 3½ years, and 5 months, respectively. Additional analysis indicated that only one of the three cases had all of the characteristics (high dose of chloramphenicol, absence of other leukemogenic agents, reasonable latency period, etc.) that suggested a causal relationship between the use of chloramphenicol and subsequent leukemia [155]. Clearly, additional studies are necessary.

TOBACCO AND ALCOHOL USE

There is no clear evidence of a strong association between the leukemias and tobacco use. Three large prospective studies have been reported. Kahn [188] reported on 8½ years of observation of smoking and mortality among 293,658 U.S. white male veterans identified from insurance policies. A relative risk of 1.49 for all leukemias (histopathologic types not specified) was found for current smokers. Veterans who were ex-smokers had a mortality ratio of 1.55. No relationship was found for pipe and cigar smokers. Hammond [189] reported the results of a 4-year follow-up of approximately one million men and women from across the United States. A relative risk for leukemia of 1.40 was found for male smokers aged 45–64 years and 1.68 for male smokers aged 65–79. Risks for leukemia in female smokers were calculated for the totality of lymphoproliferative diseases combined. A third study was a 20-year follow-up of 34,440 British male physicians who had completed questionnaires about smoking habits in 1951, 1957, 1966, and 1972. Unique features of this study were very complete follow-up (97% in the 20-year follow-up), accurate death certification, and the reduced cigarette consumption of the study population during the period of observation. No evidence of an association between smoking and leukemia was noted [190].

Data from the interview portion of the Third National Cancer Survey included an examination of the association of leukemias (and other cancers) with tobacco [191]. Relative odds were calculated by comparing patients from one cancer site to "controls" from all other sites. A suggestive dose-response relationship of smoking with AML in females and CML in both males and females was found as shown in Table 6-17, but no relationship of smoking and ALL and CLL was noted.

In the same Third National Cancer Interview Survey, the relationship with alcohol consumption was ascertained. No consistent association with leukemia was noted. Relative odds greater than 2 were noted for females with ALL (O.R. = 3.50), males with AML (O.R. = 2.56), and males with CML (O.R. = 3.26) for alcohol consumption $<$ 50 total ounce-

Table 6-17 Age- and race-adjusted association of cigarette smoking with the leukemias in males and females

Type of leukemia	Sex	20 pack years	20–40 pack years	40+ pack years
ALL	Males	0.40	1.48	0.35
	Females	1.14	—	—
CLL	Males	1.36	0.84	0.78
	Females	0.84	0.34	0.53
AML	Males	1.61	1.35	1.14
	Females	1.59	8.76	2.59
CML	Males	1.80	—	3.22
	Females	0.75	3.27	2.59

Source: Third National Cancer Survey [147].

Table 6-18 Age- and race-adjusted association of alcohol use with the leukemias in males and females

Type of leukemia	Sex	< 50 total ounce-years	≥ 50 total ounce-years
ALL	Males	—	0.88
	Females	3.50	—
CLL	Males	1.75	0.94
	Females	0.86	0.98
AML	Males	2.56	1.08
	Females	2.34	—
CML	Males	3.26	0.94
	Females	0.95	—

Source: Third National Cancer Survey [147].

years, but the significance of this was not clear since similar associations were not found in these same groups for consumption ≥ 50 total ounce-years (Table 6-18) [191].

REFERENCES

1. United Nations Scientific Committee on the Effects of Atomic Radiation, *Sources and Effects of Ionizing Radiation,* 1977 Report to the General Assembly, pp. 571–602. New York: United Nations Publication, 1977.
2. Report of the Committee on the Biological Effects of Ionizing Radiation, *The Effects of Populations of Exposure to Low Levels of Ionizing Radiation (The 1980 BEIR Report)*. Washington, D.C.: National Academy of Sciences/National Research Council, 1980.
3. G.W. Beebe, "Ionizing radiation and health," *Am. Sci.* **70**:35–44 (1982).
4. H.I. Kohn, R.J.M. Fry, "Radiation carcinogenesis," *N. Engl. J. Med.* **310**:504–509 (1984).
5. G.W. Beebe, "The atomic bomb survivors and the problem of low-dose radiation effects," *Am. J. Epidemiol.* **114**:761–783 (1981).
6. T. Ishimaru, M. Otake, and M. Ichimaru, "Dose-response relationship of neutrons and gamma rays to leukemia incidence among atomic bomb survivors in Hiroshima and Nagasaki by type of leukemia, 1950–1971," *Radiat. Res.* **77**:377–394 (1979).
7. H. Kato, W.J. Schull, "Studies of the mortality of A-bomb survivors," Report 7. Mortality, 1950–1978: Part 1. Cancer Mortality. *Radiat. Res.* **90**:395–432 (1982).

8. M. Ichimaru, T. Ishimaru, and J.L. Belsky "Incidence of leukemia in atomic bomb survivors belonging to a fixed cohort in Hiroshima and Nagasaki, 1950–71: radiation dose, years after exposure, age at exposure, and type of leukemia," *J. Rad. Res.* **19:**262–282 (1978).
9. T. Wakabayashi, H. Kato, T. Ikeda, and W.J. Schull, "Studies of the mortality of A-bomb survivors," Report 7. Part III. Incidence of cancer in 1959–1978, based on the Tumor Registry, Nagasaki. *Radiat. Res.* **93:**112–146 (1983).
10. J.A. Auxier, "Physical dose estimates for A-bomb survivor," Studies at Oak Ridge, U.S.A. *J. Radiat. Res.* (Tokyo) **16**(Suppl.):1–11 (1975).
11. W.E. Loewe and E. Mendelsohn, "Revised dose estimates at Hiroshima and Nagasaki," *Health Phys.* **41:**663–666 (1981).
12. W.K. Sinclair and P. Failla, "Dosimetry of the atomic bomb survivors," *Radiat. Res.* **88:**437–447 (1981).
13. G.D. Kerr, "A review of dosimetry for the atomic bomb survivors," in *Proceedings of the Fourth Symposium on Neutron Dosimetry*. Vol 1. (G. Burger, H. G. Ebert, eds.) Luxembourg: Munich-Neuherberg, 1981.
14. M. Ichimaru, T. Ishimaru, J.L. Belsky et al., "Incidence of leukemia in A-bomb survivors by dose, years of exposure, age, and type of leukemia, 1950–71, Hiroshima and Nagasaki," RERF Technical Report #10–76 (1976).
15. G.W. Beebe, H. Kato and C.E. Land "Mortality experience of atomic bomb survivors 1950–74." Life Span Study Report 8. RERF Technical Report #1–77 (1977).
16. S. Jablon and H. Kato, "Childhood cancer in relation to prenatal exposure to atomic bomb radiation," *Lancet* **2:**1000–1003 (1970).
17. E.S. Weiss, M.L. Rallison, W.T. London, and G.D. Thompson, "Thyroid nodularity in southwestern Utah school children exposed to fallout radiation," *Am. J. Pub. Health* **61:**241–249 (1971).
18. M.L. Rallison, B.M. Dobyns, F.R. Keating, J.E. Rall, and F.H. Tyler, "Thyroid disease in children: a survey of subjects potentially exposed to fallout radiation," *Am. J Med.* **56:**457–463 (1974).
19. M.L. Rallison, B.M. Dobyns, F.R. Keating, J.E. Rall, and F.H. Tyler, "Thyroid nodularity in children," *J. Am. Med. Assoc.* **233:**1069–1072 (1975).
20. R.A. Conard, K.D. Knudsen, and B.M. Dobyns, *A Twenty-Year Review of Medical Findings in a Marshallese Population Accidentally Exposed to Radioactive Fallout.* Upton, New York: Brookhaven National Laboratory, 1975.
21. J. Lyon, M. Klauber, J. Gardner, and K.S. Udall, "Childhood leukemias associated with fallout from nuclear testing," *N. Engl. J. Med.* **300:**397–402 (1979).

22. C.R. Stark and A. Oleinick, "Urban or rural residence and histologic type distribution in 21,000 childhood leukemia deaths in the United States, 1950–59," *J. Natl. Cancer Inst.* **37**:369–379 (1966).
23. J.L. Lyon, J. Gardner, D. West, and L. Schussman, "Further information on the association of childhood leukemias with atomic fallout," in *Cancer Incidence in Defined Population,* Banbury Series, No. 4, pp. 145–162. Cold Spring Harbor, NY: Cold Spring Harbor Laboratory, 1980.
24. C. Land, "The hazards of fallout or of epidemiological research?" *N. Engl. J. Med.* **300**:431–432 (1979).
25. Comptroller General's Report to the Congress of the United States, *Problems in Assessing the Cancer Risks of Low-Level Ionizing Radiation Exposure,* Vol. 2. Washington, D.C.: U.S. General Accounting Office, 1981.
26. J.E. Enstrom, "The nonassociation of fallout radiation with childhood leukemia in Utah," in *Cancer Incidence in Defined Populations,* Banbury Report, No. 4. (J. Cairns, J.L. Lyon, and M. Skolnick, eds). Cold Spring Harbor, NY: Cold Spring Harbor Laboratory, 1980.
27. H.L. Beck and P.W. Krey, "Radiation exposures in Utah from Nevada nuclear tests," *Science* **220**:18–24 (1983).
28. C.E. Land, F.W. McKay, and S.G. Machado, "Childhood leukemia and fallout from the Nevada nuclear tests," *Science* **223**:139–144 (1984).
29. G. Caldwell, D. Kelley, and C. Heath, Jr., "Leukemia among participants in military maneuvers at a nuclear bomb test: a preliminary report," *J. Am. Med. Assoc.* **244**:1575–1578 (1980).
30. R. Seltser and P.E. Sartwell, "The influence of occupational exposure on the mortality of American radiologists and other medical specialists," *Am. J. Epidemiol.* **81**:2–22 (1965).
31. G.M. Matanoski, R. Seltser, P.E. Sartwell, E.L. Diamond, and E.A. Elliott, "The current mortality rates of radiologists and other physician specialists: deaths from all causes and from cancer," *Am. J. Epidemiol.* **101**:188–198 (1975).
32. G.M. Matanoski, R. Seltser, P.E. Sartwell, E.L. Diamond, and E.A. Elliott, "The current mortality rates of radiologists and other physician specialists: specific causes of death," *Am. J. Epidemiol.* **101**:199–210 (1975).
33. H. Ulrich, "The incidence of leukemia in radiologists," *N. Engl. J. Med.* **234**:45–46 (1946).
34. H.C. March, "Leukemia in radiologists in a 20-year period," *Am. J. Med. Sci.* **220**:282–286 (1950).
35. S. Warren, "Longevity and causes of death from irradiation in physicians," *J. Am. Med. Assoc.* **162**:464–468 (1956).

36. S. Jablon, "Radiation," in *Persons at High Risk of Cancer: An Approach to Cancer Etiology and Control,* pp. 151–165. (J.F. Fraumeni, Jr., ed). New York: Academic Press, 1975.
37. W.M. Court-Brown and R. Doll, "Expectation of life and mortality from cancer among British radiologists," *Br. Med. J.* **2**:181–187 (1958).
38. R.W. Miller and S. Jablon, "A search for late radiation effects among men who served as x-ray technologists in the U.S. Army during World War II," *Radiology* **96**:269–274 (1970).
39. T.F. Mancuso, A. Stewart, and G. Kneale, "Radiation exposures of Hanford workers dying from cancer and other causes," *Health Phys.* **33**:369–384 (1977).
40. G.W. Kneale, T.F. Mancuso, and A.M. Stewart, "Hanford radiation study III: a cohort study of the cancer risks from radiation to workers at Hanford (1944–77 deaths) by the method of regression models in life-tables," *Br. J. Ind. Med.* **38**:156–166 (1981).
41. A. Stewart, G. Kneale, and T. Mancuso, "The Hanford data—a reply to recent criticisms," *Ambio* **9**:66–73 (1980).
42. G.B. Hutchison, B. MacMahon, S. Jablon, and C.E. Land, "Review of report by Mancuso, Stewart, and Kneale of radiation exposure of Hanford workers," *Health Phys.* **37**:207–220 (1979).
43. E.S. Gilbert and S. Marks, "An analysis of the mortality of workers in a nuclear facility," *Rad. Res.* **79**:122–148 (1979).
44. T. Anderson, "Radiation exposure of Hanford workers: a critique of the Mancuso, Stewart and Kneale report," *Health Phys.* **35**:743–750 (1978).
45. T. Najarian and T. Colton, "Mortality from leukemia and cancer in shipyard nuclear workers," *Lancet* **1**:1018–1020 (1978).
46. T. Colton, R. Greenberg, and J. Barron, "An analysis of deaths among nuclear shipyard workers," Final Report to the National Institute for Occupational Safety and Health, 1980.
47. V. Archer, J. Wagoner, and F. Lundin, "Lung cancer among uranium miners in the United States," *Health Phys.* **25**:351–371 (1973).
48. G. Saccomanno, V.E. Archer, O. Auerbach, M. Kuschner, R.P. Saunders, and M.G. Klein, "Histologic types of lung cancer among uranium miners," *Cancer* **27**:515–523 (1971).
49. A. DeVilliers and J. Windish, "Lung cancer in a fluorspar mining community: I. Radiation, dust, and mortality experience," *Br. J. Ind. Med.* **21**:94–109 (1964).
50. J.T. Boyd, R. Doll, J.S. Faulds, and J. Leiper, "Cancer of the lung in iron ore (haematite) miners," *Br. J. Ind. Med.* **27**:97–105 (1970).
51. A. Pirchan and H. Sikl, "Cancer of the lung in the miners of Jachymov (Joachimstal): report of the cases observed in 1929–30," *Am. J. Cancer* **16**:681–722 (1932).

52. J.K. Wagoner, R.W. Miller, F.E. Lundin, Jr., J.F. Fraumeni, Jr., and M.E. Haij, "Unusual cancer mortality among a group of underground metal miners," *N. Engl. J. Med.* **269**:284–289 (1963).
53. S. Graham, M. Levin, A. Lilienfeld, L. Schuman, R. Gibson, J. Dowd, and L. Hempelmann, "Preconception, intra-uterine, and postnatal irradiation as related to leukemia," *Natl. Cancer Inst. Monogr.* **19**:347–371 (1966).
54. R. Gibson, I.D.J. Bross, S. Graham, A. Lilienfeld, L.M. Schuman, M.L. Levin, and J.E. Dowd, "Leukemia in children exposed to multiple risk factors," *N. Engl. J. Med.* **279**:906–909 (1968).
55. R. Gibson, S. Graham, A. Lilienfeld, L. Schuman, J.E. Dowd, and M.L. Levin, "Irradiation in the epidemiology of leukemia among adults," *J. Natl. Cancer Inst.* **48**:301–311 (1972).
56. I. Bross and N. Natarajan, "Leukemia from low-level radiation: identification of susceptible children," *N. Engl. J. Med.* **287**:107–110 (1972).
57. I. Bross, M. Ball, and S. Falen, "A dosage response curve for the one rad range: adult risk from diagnostic radiation," *Am. J. Pub. Health* **69**:130–136 (1979).
58. J. Boice and C.E. Land, "Adult leukemia following diagnostic x-rays?" *Am. J. Pub. Health* **69**:137–145 (1979).
59. F. Gunz and H. Atkinson, "Medical radiation and leukemia: a retrospective survey," *Br. Med. J.* **1**:389–393 (1964).
60. A. Linos, J. Gray, A. Orvis, R.A. Kyle, W.M. O'Fallon, and L.T. Kurland, "Low dose radiation and leukemia," *N. Engl. J. Med.* **302**:1101–1105 (1980).
61. A. Stewart, J. Webb, and D. Hewitt, "A survey of childhood malignancies," *Br. Med. J.* **1**:1495–1508 (1958).
62. A. Stewart, "Cancer as the cause of abortion and stillbirths: the effects of these early deaths on the recognition of radiogenetic leukemias," *Br. J. Cancer* **27**:465–472 (1973).
63. J. Bithell and A. Stewart, "Prenatal irradiation and childhood malignancy: a review of British data from the Oxford survey," *Br. J. Cancer* **31**:271–287 (1975).
64. E. Diamond, H. Schmerler, and A. Lilienfeld, "The relationship of intra-uterine radiation to subsequent mortality and development of leukemia in children," *Am. J. Epidemiol.* **97**:283–313 (1973).
65. B. MacMahon, "Prenatal x-ray exposure and childhood cancer," *J. Natl. Cancer Inst.* **28**:1173–1191 (1962).
66. R.E. Albert and A.R. Omran, "Follow-up study of patients treated by x-ray epilation for tinea capitis," *Arch. Environ. Health* **17**:899–918 (1968).
67. N. Hurley, R. Albert, R. Shore, and B.S. Pasternack, "Follow-up

study of patients treated by x-ray epilation for tinea capitis. Estimation of the dose to the thyroid and pituitary glands and other structures in the head and neck," *Phys. Med. Biol.* **21**:631–642 (1976).

68. L.H. Hempelmann, J.W. Pifer, G.J. Burke, R. Terry, and W.R. Ames, "Neoplasms in persons treated with x-rays in infancy for thymic enlargement. A report of the third follow-up survey," *J. Natl. Cancer Inst.* **38**:317–341 (1967).
69. L. Hempelmann, W. Hall, M. Phillips, R.A. Cooper, and W.R. Ames, "Neoplasms in persons treated with x-ray in infancy: fourth survey in 20 years," *J. Natl. Cancer Inst.* **55**:519–530 (1975).
70. E.L. Saenger, G.E. Thomas, and E.A. Tompkins, "Incidence of leukemia following treatment of hyperthyroidism," *J. Am. Med. Assoc.* **205**:855–862 (1968).
71. E. Pochin, "Long-term hazards of radioiodine treatment of thyroid carcinoma," in *Thyroid Cancer,* UICC Monograph Series, Vol. 12, pp. 293–299 (C. Hedinger, ed). Berlin: Springer-Verlag, 1969.
72. B. Modan, D. Buidatz, and H. Mart, "Radiation-induced head and neck tumors," *Lancet* **1**:277–279 (1974).
73. B. Modan, E. Ron, and A. Werner, "Thyroid cancer following scalp irradiation," *Radiology* **123**:741–744 (1977).
74. M. Favus, A. Schneider, M. Stachura, J.E. Arnold, V.Y. Ryo, S.M. Pinsky, M. Colman, M. Arnold, and L.A. Frohman, "Thyroid cancer occurring as a late consequence of head and neck irradiation," *N. Engl. J. Med.* **294**:1019–1025 (1976).
75. R. Doll and P. Smith, "The long-term effects of x-irradiation in patients treated for metropathia haemorrhagica," *Br. J. Radiol.* **41**:362–368 (1968).
76. P. Smith and R. Doll, "Late effects of x-irradiation in patients treated for metropathia haemorrhagica," *Br. J. Radiol.* **49**:224–232 (1976).
77. M. Alderson and S. Jackson, "Long term follow-up of patients with menorrhagia treated by irradiation," *Br. J. Rad.* **44**:295–298 (1971).
78. C.L. Randall, F.P. Paloucek, J.B. Graham, and S. Graham, "Causes of death in cases of preclimacteric menorrhagia," *Am. J. Obstet. Gynecol.* **88**:880–897 (1964).
79. H. Kohn, J.C. Bailar III, and C. Zippin, "Radiation therapy for cancer of the cervix: its late effects on the lifespan as a function of regional dose," *J. Natl. Cancer Inst.* **34**:345–361 (1965).
80. C. Zippin, J.C. Bailar, H.I. Kohn, D. Lum, and H. Eisenberg, "Radiation therapy for cervical cancer: late effects on lifespan and on leukemia incidence," *Cancer* **28**:937–942 (1971).

81. G. Hutchison, "Leukemia in patients with cancer of the cervix uteri treated with radiation. A report covering the first 5 years of an international study," *J. Natl. Cancer Inst.* **40:**951–982 (1968).
82. M. Sulzberger, R.L. Baer, and A. Borota, "Do roentgen-ray treatments as given by skin specialists produce cancers or other sequelae?" *Arch. Dermatol. Syph.* **65:**639–655 (1952).
83. J. da Silva Horta, L. Cayolla da Motta, and M.H. Tavares, "Thorium dioxide effects in man," *Environ. Res.* **8:**131–159 (1974).
84. J. da Silva Horta, M.E. da Silva Horta, L. Cayolla da Motta, and M.H. Tavares, "Malignancies in Portugese thorotrast patients," *Health Phys.* **35:**137–151 (1978).
85. B. Modan and A.M. Lilienfeld, "Polycythemia vera and leukemia—the role of radiation treatment: a study of 1,222 patients," *Medicine* **44:**305–344 (1965).
86. W.M. Court-Brown and R.L. Doll, "Leukemia and aplastic anemia in patients irradiated for ankylosing spondylitis," *Med. Res. Counc.* (G.B.) *Spec. Rep. Ser.* **295** (1957).
87. W.M. Court-Brown and R. Doll, "Mortality from cancer and other causes after radiotherapy for ankylosing spondylitis," *Br. Med. J.* **2:**1327–1332 (1965).
88. E.P. Radford, R. Doll, and P.G. Smith, "Mortality among patients with ankylosing spondylitis not given x-ray therapy," *N. Engl. J. Med.* **297:**572–576 (1977).
89. J. Higginson, "Present trends in cancer epidemiology," *Proc. Can. Cancer Conf.* **8:**40–75 (1969).
90. E.L. Wynder and G.B. Gori, "Contribution of the environment to cancer incidence," *J. Natl. Cancer Inst.* **58:**825–832 (1977).
91. P.E. Enterline, "Extrapolation from occupational studies: a substitute for environmental epidemiology," *Environ. Health Perspect.* **42:**39–44 (1981).
92. Carcinogen Assessment Group's Final Report on Population Risk to Ambient Benzene Exposures. Washington, D.C.: Environmental Protection Agency, 1978.
93. P. Delore and Borgomano, "Leucémie aiguë au cours de l'intoxication benzénique," *J. Med. (Lyon)* **9:**227–235 (1928).
94. R. Penati and E.C. Vigliani, "Sul problema delle mi elopathie asplastiche pseudo asplastiche e leucemiche da benzolo," *Rass. Med. Ind.* **9:**345–361 (1938).
95. T.B. Mallory, E.A. Gall, and S.J. Brickley, "Chronic exposure to benzene (benzol). III. The pathologic results," *J. Ind. Hyg. Toxicol.* **21:**355–377 (1939).
96. E. Browning, "Vitamins in the treatment of chronic benzene poisoning," *J. Occup. Med.* **7:**554–559 (1965).

97. E.C. Vigliani and A. Forni, "Benzene and leukemia," *Environ. Res.* **11**:122–127 (1976).
98. E.C. Vigliani, "Leukemia associated with benzene exposure," *Ann. N.Y. Acad. Sci.* **271**:143–151 (1976).
99. L. Brandt, P.G. Nilsson, and F. Mitelman, "Non-industrial exposure to benzene as a leukaemogenic risk factor," *Lancet* (Letter to the Editor) **2**:1074 (1977).
100. L. Brandt, P.G. Nilsson, and F. Mitelman, "Occupational exposure to petroleum products in men with acute non-lymphocytic leukemia," *Br. Med. J.* **1**:553 (1978).
101. F. Mitelman, L. Brandt, and P.G. Nilsson, "Relation among occupational exposure to potential mutagenic/carcinogenic agents, clinical findings, and bone marrow chromosomes in acute nonlymphocytic leukemia," *Blood* **52**:1229–1237 (1978).
102. M. Aksoy, S. Erdem, and G. Dincol, "Leukemia in shoe-workers exposed chronically to benzene," *Blood* **44**:837–841 (1974).
103. M. Aksoy, S. Erdem, and G. Dincol, "Types of leukemia in a chronic benzene poisoning: a study in thirty-four patients," *Acta Haematol.* **55**:65–72 (1976).
104. P.F. Infante, J.K. Wagoner, R.A. Rinsky, and R.J. Young, "Leukaemia in benzene workers," *Lancet* **2**:76–78 (1977).
105. 43 Federal Register 5918, "Preamble to the rule regulating exposure to benzene," Department of Labor, O.S.H.A., 1978.
106. M.G. Ott, J.C. Townsend, W.A. Fishback, and F.A. Langer, "Mortality among individuals occupationally exposed to benzene," *Arch. Environ. Health* **33**:3–10 (1978).
107. J.J. Thorp, "Epidemiologic survey of leukemia in persons potentially exposed to benzene," JOM **16**:375–382 (1974).
108. Tabershaw Cooper, Inc., "A morbidity study of petroleum refinery workers," American Petroleum Institute, Washington, D.C., 1975.
109. T. Ishimaru, H. Okada, T. Tomiyasu, and T. Tsuchimoto, "Occupational factors in the epidemiology of leukemia in Hiroshima and Nagasaki," *Am. J. Epidemiol.* **93**:157–165 (1971).
110. V.L. Jedlicka, Z. Hermanska, I. Smida, and A. Kouba, "Paramyeloblastic leukaemia appearing simultaneously in two blood cousins after simultaneous contact with Gammexane (hexachlorcyclohexane)," *Acta Med. Scand.* **161**:447–451 (1958).
111. I. West, "Lindane and hematologic reactions," *Arch. Environ. Health* **15**:97–101 (1967).
112. H. Hoshizaki, Y. Niki, H. Tajima, Y. Terada, and A. Kasahara, "A case of leukemia following exposure to insecticide," *Acta Haematol. Jpn.* **32**:672–677 (1969).
113. T.T. Timonen and I.P. Palva, "Acute leukaemia after exposure to a

weed killer, 2-methyl-4-chlorphenoxyacetic acid," *Acta Haematol. (Basel)* **63:**170–171 (1980).

114. W.A. Blattner, W. Strober, A.V. Muchmore, R.M. Blaese, S. Broder, and J.F. Fraumeni, Jr., "Familial chronic lymphocytic leukemia," *Ann. Intern. Med.* **84:**554–557 (1976).
115. F.A. Walker, "Familial leukemia and dry cleaning," *Ann. Intern. Med.* **85:**404–405 (1976).
116. A.J. McMichael, R. Spirtas, and L.L. Kupper, "An epidemiologic study of mortality within a cohort of rubber workers, 1964–72," JOM **16:**458–459 (1974).
117. A.J. McMichael, R. Spirtas, L.L. Kupper, and J.F. Gamble, "Solvent exposure and leukemia among rubber workers: an epidemiologic study," JOM **17:**234–239 (1975).
118. A.J. McMichael, R. Spirtas, J.F. Gamble, and P.M. Tousey, "Mortality among rubber workers: relationship to specific jobs," JOM **18:**178–185 (1976).
119. R.H. Monson, and K.K. Nakano, "Mortality among rubber workers," *Am. J. Epidemiol.* **103:**284–296 (1976).
120. J. Fabia and T.D. Thuy, "Occupation of father at time of birth of children dying of malignant diseases," *Br. J. Prev. Soc. Med.* **28:**98–100 (1974).
121. T. Hakulinen, T. Salonen, and L. Teppo, "Cancer in the offspring of fathers in the hydrocarbon-related occupations," *Br. J. Prev. Soc. Med.* **30:**138–140 (1976).
122. M. Zack, S. Cannon, D. Lloyd, C.W. Heath, Jr., J.M. Falletta, B. Jones, J. Housworth, and S. Crowley, "Cancer in children of parents exposed to hydrocarbon-related industries and occupations," *Am. J. Epidemiol.* **111:**329–336 (1980).
123. R. Olin, "Leukemia and Hodgkin's disease among Swedish chemistry graduates," *Lancet* (Letter to the Editor) **2:**916 (1976).
124. T. Sugiyama, H. Venaka, N. Veda, S. Fukuhara, and S. Maeda, "Reproducible chromosomal changes of polycyclic hydrocarbon-induced rat leukemia: incidence and chromosome banding pattern," *J. Natl. Cancer Inst.* **60:**153–160 (1978).
125. A. Ducatman, K. Hirschhorn, and I.J. Selikoff, "Vinyl chloride exposure and human chromosome aberrations," *Mutat. Res.* **31:**163–168 (1975).
126. I.M. Tough, P.G. Smith, W.M. Court-Brown, and D.G. Harnden, "Chromosome studies on workers exposed to atmospheric benzene: the possible influence of age," *Eur. J. Cancer* **6:**49–55 (1970).
127. I.F.H. Purchase, C. Richardson, and D. Anderson, "Chromosomal effects in peripheral lymphocytes," *Proc. R. Soc. Med.* **69:**290–291 (1976).

128. F. Funes-Cravioto, C. Zapata-Gayon, and B. Kolmodian-Hedman, "Chromosome aberrations and sister chromatid exchange in workers in chemical laboratories and a rotoprinting factory and in children of women laboratory workers," *Lancet* **2**:322–325 (1977).
129. A.M. Forni, E. Pacifido, and A. Limonta, "Chromosome studies in workers exposed to benzene or toluene or both," *Arch. Environ. Health* **22**:373–378 (1971).
130. S.S. Epstein, *The Politics of Cancer,* p. 133. Garden City, N.Y.: Anchor Press, 1979.
131. J. Lieben, "Malignancies in asbestos workers," *Arch. Environ. Health* **13**:619–621 (1966).
132. M.A. Gerber, "Asbestosis and neoplastic disorders of the hematopoietic system," *Am. J. Clin. Pathol.* **53**:204–208 (1970).
133. E. Fasal, E.W. Jackson, and M.R. Klauber, "Leukemia and lymphoma mortality and farm residence," *Am. J. Epidemiol.* **87**:267–274 (1968).
134. S. Milham, Jr., "Leukemia and multiple myeloma in farmers," *Am. J. Epidemiol.* **94**:307–310 (1971).
135. A. Blair and T.L. Thomas, "Leukemia among Nebraska farmers: a death certificate study," *Am. J. Epidemiol.* **110**:264–273 (1979).
136. S. Milham, Jr., "Mortality from leukemia in workers exposed to electrical and magnetic fields," *N. Engl. J. Med.* (Letter to the Editor) **307**:249 (1982).
137. W.E. Wright, J.M. Peters, and T.M. Mack, "Leukaemia in workers exposed to electrical and magnetic fields," *Lancet* (Letter to the Editor) **2**:1160–1161 (1982).
138. M.E. McDowall, "Leukemia mortality in electrical workers in England and Wales," *Lancet* (Letter to the Editor) **1**:246 (1983).
139. M. Coleman, J. Bell and R. Skeet, "Leukaemia incidence in electrical workers," *Lancet* **1**:982–983 (1983).
140. Editorial, "Electromagnetism and cancer," *Lancet* **1**:224 (1983).
141. N. Wertheimer and E. Leeper, "Electrical wiring configurations and childhood cancer," *Am. J. Epidemiol.* **109**:273–284 (1979).
142. J.P. Fulton, S. Cobb, L. Preble, L. Leone and E. Forman, "Electrical wiring configurations and childhood leukemia in Rhode Island," *Am. J. Epidemiol.* **111**:292–296 (1980).
143. E. Viadana and I.D.J. Bross, "Leukemia and occupations," *Prev. Med.* **1**:513–521 (1972).
144. Registrar General, *Registrar General's Decennial Supplement England and Wales, 1971. Occupational Mortality,* London: Her Majesty's Stationery Office, 1978.
145. P. Decoufle, K. Stanislawczyk, L. Houten, I.D.J. Bross, and E. Via-

dana, *A Retrospective Survey of Cancer in Relation to Occupation,* U.S.D.H.E.W., P.H.S., C.D.C., N.I.O.S.H., Division of Surveillance, Hazard Evaluations, and Field Studies, Pub. No. 77-178. Washington, D.C.: U.S. Government Printing Office, 1977.

146. S. Milham, *Occupational Mortality in Washington State, 1950–71,* Vols. 1–3 U.S.D.H.E.W., P.H.S. Washington, D.C.: U.S. Government Printing Office, 1976.
147. R.G. Williams, N.L. Stegens, and J.R. Goldsmith, "Associations of cancer site and type with occupation and industry from The Third National Cancer Survey Interview,"*J. Natl. Cancer Inst.* **59:**1147–1185 (1977).
148. R. Doll, "Statistical approach to the evaluation and prediction of possible carcinogenic action of drugs," in *Potential Carcinogenic Hazards from Drugs,* pp. 53–58 (R. Truhart, ed), New York: Springer-Verlag, 1967.
149. I. Berenblum, "Principles of methodology relating to the potential carcinogenicity of medical drugs," in *Potential Carcinogenic Hazards from Drugs,* pp. 28–32 (R. Truhart, ed). New York: Springer-Verlag, 1967.
150. D. Schmähl, "Carcinogenic hazards from drugs," in *Cancer Detection and Prevention,* International Cancer Series No. 322, pp. 34–40 (C. Maltoni, ed). Amsterdam: Excerpta Medica, 1973.
151. J. Juhasz, J. Balo, and G. Kendrey, "Uber die geschwalsterzengende Wirkung des Isonikotinsaurehydrazid (INH)," *Z. Krebsforsch.* **62:**188–191 (1957).
152. S. Schwan, "Hydrazid kwasu izonikotynowego (INH) jako czynnik 'rakotworczy' u myszy," *Doniesienie II. Patol. Pol.* **13:**185–189 (1962).
153. V. Milia, M. Gaetani, and C. e Biancifiori, "Azione carcinogenetica dell'acido 4-(isonicotinil-idrazone) pimelico (4-INIP) in topi BALB/c/cb/Se subtrain," *Lav. Ist. Anat. Istol. Patol. Univ. Studi Perugia* **24:**39–43 (1964).
154. J.N. Berry, "Acute myeloblastic leukemia in a Benzedrine® addict," *South. Med. J.* **59:**1169–1170 (1966).
155. H. Welch, C.N. Lewis, and I. Kerlan, "Blood dyscrasias. A nationwide survey," *Antiobiot. Chemother.* **4:**607–623 (1954).
156. A.M. Davies, B. Modan, M. Djaldetti, and A. deVries, "Epidemiological observations on leukemia in Israel," *Arch. Intern. Med.* **108:**86–90 (1961).
157. T. Cohen and W.P. Creger, "Acute myeloid leukemia following seven years of aplastic anemia induced by chloramphenicol," *Am. J. Med.* **43:**762–770 (1967).

158. J.F. Fraumeni, Jr., "Bone marrow depression induced by chloramphenicol or phenylbutazone. Leukemia and other sequelae," *J. Am. Med. Assoc.* **201**:828–834 (1967).
159. M.J. Brauer and W. Dameshek, "Hypoplastic anemia and myeloblastic leukemia following chloramphenicol therapy," *N. Engl. J. Med.* **277**:1003–1005 (1967).
160. K.R. Humphries, "Acute myelomonocytic leukemia following chloramphenicol therapy," *N. Z. Med. J.* **68**:248–249 (1968).
161. K. Goh, "Chloramphenicol, acute leukemia and chromosomal vaculizations," *South. Med. J.* **64**:815–819 (1971).
162. H.J. Cohen and A. Huang, "A marked chromosome abnormality. Occurrence in chloramphenicol-associated acute leukemia," *Arch. Intern. Med.* **132**:440–443 (1973).
163. P.A. Cassileth, "Monocytosis in chlorpromazine-associated agranulocytosis. Termination in acute leukemia," *Am. J. Med.* **43**:471–476 (1967).
164. E. König, K. Berthold, H.A. Hienz, and G. Brittinger, "Griseofulvin and chronic granulocytic leukaemia," *Helv. Med. Acta* **35**:103–107 (1969).
165. L. Grossbard, E. McGilvray, and O. Miller, "Acute leukemia with Ph^1-like chromosome in an LSD user," *J. Am. Med. Assoc.* **205**:791–793 (1968).
166. O.M. Garson and M.K. Robson, "Studies in a patient with acute leukemia after lysergide treatment," *Br. Med. J.* **2**:800–802 (1969).
167. S.G. Sjoberg and D. Perers, "Leukemia after oxyphenbutazone," *Lancet* (Letter to the Editor) **2**:441 (1965).
168. D. Perers, "Akut leukemia, leukemoid reaktion och leukocytos after behandling med oxifenylbutazone," *Sven Laekartidn.* **63**:53 (1966).
169. R.H.D. Bean, "Phenylbutazone and leukaemia," *Br. Med. J.* **2**:1552–1555 (1960).
170. E.F.B. Cadman and W. Limont, "Phenylbutazone and leukemia," *Br. Med. J.* (Letter to the Editor) **1**:798 (1962).
171. H.J. Woodliff and L. Dougan, "Acute leukaemia associated with phenylbutazone treatment," *Br. Med. J.* **1**:744–746 (1964).
172. T.M. Chalmers and D.D. McCarthy, "Phenylbutazone therapy associated with leukemia," *Br. Med. J.* **1**:747 (1964).
173. D. McCarthy and T.M. Chalmers, "Hematological complications of phenylbutazone therapy: review of the literature and report of two cases," *Can. Med. Assoc. J.* **90**:1061–1067 (1964).
174. L. Dougan and H.J. Woodliff, "Acute leukemia associated with phenylbutazone treatment: a review of the literature and report of a further case," *Med. J. Aust.* **1**:217–219 (1965).

175. G.M. Leavesley, N.S. Stenhouse, L. Dougan, and H.J. Woodliff, "Phenylbutazone and leukemia. Is there a relationship?" *Med. J. Aust.* **2**:963–965 (1969).
176. M. Aksoy, S. Erdem, H. Tezel, and T. Tezel, "Acute myeloblastic leukaemia after propylthiouracil," *Lancet* (Letter to the Editor) **1**:928–929 (1974).
177. R.C. Brod, "Blood dyscrasias associated with tolbutamide therapy," *J. Am. Med. Assoc.* **171**:296–297 (1959).
178. H. Schneider, R. Slowik, and H.W. Wache, "Monozyter leukose unter tolbutamide-lang zeittherapie," *Z. Gesamte Inn. Med. Ihre Grenzgeb.* **25**:88–90 (1970).
179. J.M. Frisch, "Clinical experience with adverse reactions to trimethoprim-sulfamethoxazole," *J. Infect. Dis.* **128**(Suppl):607–612 (1973).
180. M. Swanson and R. Cook, *Drugs, Chemicals and Blood Dyscrasias.* Hamilton, IL: Drug Intelligence Publications, 1977.
181. J.B. Chatterjea, "Leukemia and phenylbutazone," *Br. Med. J.* (Letter to the Editor) **2**:875 (1964).
182. G.J. Thorpe, "Leukaemia and phenylbutazone," *Br. Med. J.* **1**:1707 (1964).
183. A.C. Stevenson, J. Bedford, J.C.S. Hill, and H.F.H. Hill, "Chromosomal studies in patients taking phenylbutazone," *Ann. Rheum. Dis.* **30**:487–500 (1971).
184. A.C. Stevenson, P.S. Hastie, and R.K. Archer, "Chromosome findings in horses treated with phenylbutazone," *Equine Vet. J.* **4**:214–217 (1972).
185. S. Walker, D.A.P. Evans, P.A. Benn, T.R. Littler, and L.D.C. Halliday, "Phenylbutazone and chromosomal damage," *Ann. Rheum. Dis.* **34**:409–415 (1975).
186. G.D. Friedman and H.K. Ury, "Initial screening for carcinogenicity of commonly used drugs," JNCI **65**:723–733 (1980).
187. G.D. Friedman, "Phenylbutazone, musculoskeletal disease, and leukemia," *J. Chron. Dis.* **35**:233–243 (1982).
188. H.A. Kahn, "The Dorn study of smoking and mortality among U.S. veterans: report on eight and one-half years of observation," in *Epidemiological Approaches to the Study of Cancer and Chronic Diseases,* NCI Monograph **19** (W. Haenszel, ed). Bethesda, Maryland: U.S.D.H.E.W., 1966.
189. E.C. Hammond and L. Garfinkel, "The influence of health on smoking habits." In *Epidemiological Approaches to the Study of Cancer and Other Chronic Diseases,* NCI Monograph **19** (W. Haenszel, ed). Bethesda, Maryland: U.S.D.H.E.W., 1966.

190. R. Doll and R. Peto, "Mortality in relation to smoking: 20 years' observations on male British doctors," *Br. Med. J.* **2:**1525–1536 (1976).

191. R.R. Williams and J.W. Horm, "Association of cancer sites with tobacco and alcohol consumption and socioeconomic status of patients: interview study from the Third National Cancer Survey," *J. Natl. Cancer Inst.* **58:**525–547 (1977).

7 Infectious Agents

LEUKEMOGENIC VIRUSES IN ANIMALS AND MAN

Since 1908, when Ellerman and Bang first transmitted leukemia to previously healthy chickens using cell-free filtered extracts from infected chickens [1], viruses have been found to cause several types of hematopoietic and solid tumors in a wide variety of species [2]. Infectious agents, particularly viruses, have long been suspected as a cause of human leukemias. A number of methodologic approaches, primarily epidemiologic but also viral and immunologic, have been used to try to demonstrate an infectious origin for the leukemias. Despite extensive effort, a viral origin for human leukemias has not been unequivocally proven. However, several studies described below and recent reports from Gallo et al., Hinuma et al., Miyoshi et al., and Catovsky et al., identifying and characterizing human retroviruses in certain rare T-cell lymphoproliferative malignancies are suggestive [3–7]. Although there are two major groups of retroviruses, an endogenous type transmitted genetically and an exogenous type spread by infection, only the latter type is associated with the leukemias and lymphomas. The exogenous retroviruses consist of two types: one that is able to undergo viral replication and the other that has recombined with host DNA and is mostly replication-defective. The replicating type does not directly transform cells; examples include bovine leukosis virus and feline leukemia virus described more fully below. The replication-defective types of retroviruses are usually only able to replicate in the presence of competent helper viruses. The host cell sequences also known as "onc" genes that have recombined with these defective viruses allow the viruses to transform cells in vitro. An example of this type of virus is the Abelson murine leukemia virus. In chickens and cattle, the lymphoproliferative tumors caused by retroviruses are primarily of B-lymphocyte origin, but in cats and mice they are primarily T-lymphocyte cancers. Most of the B-cell and T-cell tumors are monoclonal [8, 9].

Horizontally transmitted retroviruses have been shown to produce leukemia and lymphoma in cows, cats, chickens, and gibbons [2]. There are two major forms of bovine leukemia, enzootic and sporadic [10]. The enzootic type occurs in adult cattle, is transmitted both horizontally (80+ % of transmission) and vertically (14%–18% is transplacental), and is caused by the C-type bovine leukosis virus (BLV). Sporadic bovine leukemia or leukosis is of unknown etiology but occurs in several forms: a juvenile multicentric disease, an adolescent thymic type, and a cutaneous form affecting cattle of all ages. The term "leukosis" is defined as a proliferation of leukocyte-forming tissue that includes both myelosis and lymphadenosis. Such proliferation forms the basis of leukemia. Bovine, avian, etc., leukoses are groups of diseases in cows, chickens, etc., characterized by proliferation of immature myeloid or lymphoid cells. The incubation period is long, sometimes up to seven years [11, 12]. Transmission may occur through blood, feces, urine, amniotic fluid, milk, conjunctival fluids, nasal secretions, and possibly by means of biting insects [12–14]. However, transmission usually occurs only after extensive intimate contact between animals [10, 15]. The prevalence was reported to be 17.51 per 100,000 in cattle and 2.36 per 100,000 in calves in one study [16].

Leukemia in cats is caused by the horizontally transmitted C-type feline leukemia virus (FeLV). The infection is clinically apparent in less than half of the cats who are followed soon after laboratory inoculation; a small percent of infected cats develop leukemia months to years later [17]. Although FeLV infection is common in free-roaming cats, it is quite uncommon in cats kept in isolation [18]. Similar to tumors occurring in bovine leukemia, one-third of the feline leukemia tumors do not contain either infectious virus or viral genomes. When a cat is infected, several independent biological processes may occur, two of the most important being viral replication and viral oncogenesis. Resistance to each of these processes is related to antibody production. Most cats older than four months at the time of experimental or natural infection produce both types of antibodies following acute infection and no complications result from the FeLV infection. However, cats exposed to high initial doses of virus and cats infected when they are young kittens are at high risk of persistent viremia and often develop tumors [19]. The period between infection and development of malignancy can vary from 3 months to 5 years [20]. FeLV has been found to occur in approximately 50% of free-roaming cats [18].

Avian leukosis/sarcoma viruses are C-type RNA tumor viruses which can induce a variety of malignancies, the most common of which is lymphoid leukosis. The virus can be transmitted by means of saliva, droppings, etc. As with bovine leukemia, avian virus horizontal spread requires intimate contact. Vertical transmission from the female parent is also a common infection route. Young birds infected from birth may not develop antibodies and may become persistently viremic [21–23]. Avian leukosis virus infection is highly prevalent in chickens, with antibodies in 30%–90% of chickens [24].

Primates can also be infected with C-type viruses that cause leukemia, as well as lymphoma and sarcoma [25–30]. Herpes virus saimiri has been reported to cause lymphocytic leukemia in owl monkeys and marmosets. Approximately 10 days after inoculation with virus, lymphoreticular proliferation becomes apparent. Death from leukemia occurs fom 35 to 116 days following inoculation. Although inclusion bodies, viral particles, or antigens could not be detected in fresh tumor tissue, the herpes virus saimiri genome could be demonstrated consistently after tumor tissue was cultured in vitro. Fluorescing and neutralizing antibodies first appeared in serum approximately 2–6 weeks after inoculation, around the same time that the virus could be recovered from circulating leukocytes [25, 27, 28]. DePaoli et al. reported five naturally occurring cases of granulocytic leukemia (similar to CGL) in a 195-member gibbon colony. One colony member developed granulocytic leukemia following inoculation of bone marrow from a leukemic gibbon. All had been inoculated with *Plasmodium falciparum*-infected blood although none developed clinical malaria. In addition, all had been exposed to an epizootic of *Herpes hominis* and all had been inoculated several times with dengue viruses [30]. Theilen et al. reported the occurrence of multiple fibrosarcomas and myelofibrosis with myeloid metaplasia (which may be related to ANLL and possibly CGL) in a wooley monkey believed to be due to a C-type virus, morphologically similar to those associated with the leukemia-sarcoma complex of chickens, mice, and cats [26].

In addition to horizontal spread between members of the same species, several of the leukemia viruses have been shown to cross species barriers and cause neoplasms in other species. The bovine leukemia virus is also infective for sheep and goats [31]. The etiologic agent for cattle lymphosarcoma has caused lymphosarcoma in sheep [32]. Erythroleukemia developed in chimpanzees who were fed with milk from BLV-infected cows [33].

Many anecdotal reports and analytic studies have examined possible transmission of the animal leukemia viruses to humans. Most describe leukemia cases (usually acute or chronic lymphocytic but, in one report, seven cases of ANLL) that occurred over a short time period in individuals in close contact with cows with bovine leukosis [34–39]. These reports have provided the impetus for additional, more analytic investigations but do not constitute proof of an association or causal relationship.

The U.S. Third National Cancer Survey [40] and international incidence and mortality data collected by the World Health Organization [41] have shown that some types of leukemia, particularly ALL and CLL, appear to have a higher incidence in rural areas than in urban areas. Several methodologic approaches have been used to examine possible causes for the increased incidence.

Ecologic studies have examined correlations between mortality or incidence of the leukemias and rate of leukemia or leukosis in cows in various geographic areas. Studies carried out in Sweden, Denmark, the Soviet Union, and Germany have shown high mortality or incidence rates for leukemia, not cell-specific in most studies [37, 42, 43], and for chronic lymphatic leukemia, lymphatic leukemia (not specified as acute or chronic), and acute blast cell leukemia in one study [44]. Other ecologic studies have failed to demonstrate any correlation [45, 46].

Ecologic correlations or the absence of such correlation can be difficult to interpret [47]. Since leukemias are very rare and most studies have very few cases, the addition of even one or two cases could result in a positive correlation, whereas no correlation may exist without those one or two cases. Most studies group all leukemias together; separate examination by histopathologic type could result in a correlation where none existed when all were grouped together or separate consideration could further decrease the already small number of cases to the point where no correlation could be shown (even if one existed) because of small numbers. In many studies, the actual proportion of infected cows or exposed persons is not known (or not stated). The proportion of persons at risk could be very small compared to the rest of the population and the association may not be shown because the smallest detectable relative risk could be very large. Many times, a study does not adjust for cattle herd size; the probability of animals developing leukosis within a herd may be related to herd size [48]. Many studies fail to allow for a latency period between exposure to BLV and development of leukemia in humans.

Another type of ecologic study comparing leukemia mortality and/or incidence between farmers and nonfarmers assumes that farmers are more likely to be exposed to cows with leukemia than are nonfarmers and employs a case-control approach. Three studies have shown a higher risk for leukemia in farmers, compared with nonfarmers [49–51]; two used vital records as a database [49, 50], and the third consisted of interviews of cases and controls [51]. A fourth study found no leukemia excess but did note increased incidence and mortality rates for Hodgkin's disease and multiple myeloma at younger ages in male farm workers [52].

Another method of attempting to determine whether there is an association between leukemia and exposure to leukotic cattle uses cases with leukemia compared to controls without leukemia. The two groups are compared for frequency of exposure to farming [53–55], occupational, or household exposure to cattle or poultry [37, 42–44], bovine leukosis [37–57], and feline leukemia [58–64].

In a study of incident leukemia cases in Olmstead County, Minnesota, 1965–74, Linos et al. found that the relative risk for farming among cases compared to expected was 2.9 [53]. Milham compared occupations from death certificates of leukemics with matched controls in Washington and Oregon. Leukemia, particularly lymphatic and acute leukemia in men under 60 at time of death, and multiple myeloma were found to be significantly associated with farming as compared to controls [54]. Blair and Thomas found an association between farming and leukemia (all types, O.R. = 1.25), particularly for farmers less than 60 years old at time of death from leukemia (O.R. = 1.83) [55].

Aleksandrowicz found that cattle raised by persons with CML and lymphatic leukemia had significantly higher rates of preleukemia than did cattle raised by healthy controls. He also compared the frequency of contact with cattle in 65 male leukemics with that of controls and noted that the leukemics were 3 times more likely to be exposed [37]. Milham noted that the excess of leukemia in farmers compared to controls was primarily related to an excess in chicken farmers [54]. Blair and Thomas found an increased risk of leukemia among farmers in counties with high cattle production (O.R. = 1.72) [55]. Donham et al. noted an association for males (under age 20 and over age 60) with ALL and cattle density in Iowa; in counties with excessive ALL, these investigators noted a high frequency of herds affected with bovine lymphosarcoma. A CLL excess was also found with an urban-to-rural gradient of increasing risk for males and a less-pronounced gradient for females. However, CLL was not associated with high animal density [56].

Although Aleksandrowicz noted a higher frequency of preleukemia (defined as ≥ 85% lymphocytosis) in cattle raised by persons with CLL and lymphatic leukemia compared with cattle raised by controls [37], Kvarnfors et al. found no difference in the frequency of occurrence of bovine leukosis in cattle belonging to households with one or more cases of human leukemia compared with households with no cases [57].

Case-control studies comparing history of illness and/or malignancy in cats in households have produced mixed results. Bross et al. analyzed data on the frequency of sick cats belonging to households of 1,400 leukemics with that in households of controls [58, 59]. The results, analyzed by cell type and age group, are shown below in Table 7-1. As can be seen, age-adjusted relative risks are 2 or greater for each cell type, although the association is significant at the 5% level for CLL only.

A positive history of a sick cat obtained by interview is not necessarily synonymous with the presence of FeLV or of a malignancy in the affected cats. In the Bross et al. study, information was obtained by interview of

Table 7-1 Age-specific and age-adjusted relative risks for exposure to sick cats by leukemia types

Type of leukemia	15–44	45–64	65+	Age-adjusted relative risk	p value
ALL	10.57	0.00	3.91	4.94	0.09
AML	3.30	1.10	0.98	1.97	0.74
CLL	0.00	2.13	3.41	2.62	0.03
CML	5.29	0.72	1.02	2.06	0.99

Source: Tristate Leukemia Survey [59].

cases and controls and neither serosurvey of cats nor examination of veterinarian medical records was carried out; reports of sick cats were not validated. Other studies have not shown an association between feline leukemia virus in cats and leukemia in persons living with or otherwise in close contact with the infected cats [60–63].

The results of the case-control studies thus far summarized must be interpreted cautiously. As with the ecologic approach, many of the case control studies fail to consider a latency period between exposure and disease. Since the latency period is unknown, cases and controls questioned about an exposure (such as farming, presence of cattle and poultry in close proximity, occurrence of bovine leukosis or cat leukemia) may not be asked about the status of the exposure at the appropriate interval. Recall bias may encourage cases to remember illness in household or farm animals more readily than controls. A small sample size may be a limiting factor in a number of the studies described (including the reports by Aleksandrowicz [37] and Linos et al. [53]). Many of the studies did not include validation measures, such as serologic confirmation of BLV or FeLV, which might result in misclassification. Negative histories of malignancy in cows, poultry, or cats to whom cases or controls are exposed may pose problems; negative clinical history is *not* synonymous with negative carrier status for BLV, avian leukosis virus, or FeLV [11, 63]. Nevertheless, a significant proportion of the case-control studies appear to support an association between leukemia and exposure to cattle and poultry; the evidence is somewhat less convincing for associations between leukemia and exposure to cats with feline leukemia.

Seroepidemiologic studies comparing leukemia cases with controls for frequency of various antibody titers to different viruses provide another type of indirect evidence for substantiating or disproving the infectious origin theory. Izui et al. compared 175 sera from patients with AML, CML, and ALL with those of controls for presence of antibodies against single-stranded and double-stranded DNA. The levels of single-stranded DNA binding activity were higher in all three types of leukemia compared to those of healthy controls. None of the sera exhibited a positive reaction with double-stranded DNA [64]. Anti-double-stranded DNA antibodies are seen mainly in patients with systemic lupus erythematosus while anti-single-stranded antibodies are frequently found in other diseases whose infectious agents may be directly or indirectly implicated in the triggering of the immune response to DNA [65–68].

Case-control seroepidemiologic studies could provide evidence of associations between viruses and leukemia but etiologic interpretations should be treated cautiously. Blood analyzed for antibody status from a diagnosed leukemic might be seropositive as a result of the occurrence of a viral infection following the onset of leukemia.

Gahrton et al. examined sera from 52 children with acute leukemia for antibodies to various herpes viruses including Epstein–Barr virus (EBV), cytomegalovirus (CMV), and herpes simplex as well as to other common viruses (including measles, rubella, mumps, adenovirus) [69]. The proportion of sera positive for herpes virus antibodies was lower among children ages 1–8 with ALL than among sera from age-matched children without leukemia. The difference was particularly striking for children under 4 years of age. The proportion of sera positive for antibodies to measles and rubella virus did not differ from the expected frequencies in these age groups. Lower-than-expected proportions of sera positive for mumps and adenovirus antibodies in children with leukemia were difficult to interpret because these viral infections occur with cyclic periodicity.

Levine et al. found a significantly higher geometric mean antibody titer (GMT) against EBV in a group of 34 patients with CLL (GMT was 1:444) than in a group of 20 patients with CML (GMT of 1:84) or with normal controls [70]. These investigators followed the CLL and CML patients for up to 5 years and no significant antibody changes were detected.

Certain herpes viruses have been causally implicated in cancers other than leukemia. Epstein–Barr virus has been associated with both Burkitt's lymphoma and nasopharyngeal lymphoma [71], and herpes simplex type II with cervical cancer [72]. Although virologic and immunologic methods and epidemiologic approaches have greatly advanced understanding of the role of viruses in carcinogenesis, a number of methodologic problems still remain.

Current laboratory techniques may not be able to detect the presence of a virus or antibody in an animal with a malignancy which, however, does not mean that the animal is not infected. Cell-free homogenates of tumor tissue have been shown in some instances to transmit tumor virus to inoculated animals. In fact, as Francis and Essex pointed out, the animals with viral-induced tumor malignancy have developed that malignancy because of an inability to respond immunologically, i.e., producing antibodies to the virus [20]. Failure to detect virus or viral antibodies can occur for a number of reasons. Since steps in the sequence of viral oncogenesis have not been completely defined, detection of virus and/or antibody may only be possible at certain stages in a lengthy process of oncogenesis; or the virus may be so completely incorporated into a genome that current methods cannot differentiate its DNA sequence from that of the host; or levels of antibody may be so low as to escape detection. In spite of proximity between humans and animals with viral-induced tumors, virus may not have been transmitted because contact between the two groups was not intimate and/or prolonged enough for infection to have developed in the humans. Evidence indicates that very close contact is required for horizontal transmission of bovine leukemia virus and avian leukosis virus to occur.

A number of reports have speculated that certain viral disorders, particularly infectious mononucleosis, are precursors of leukemia [73–78]. One group of investigators reported a B-cell lymphoma which appeared simultaneously with a primary infection with EBV in a child with multiple congenital anomalies [79]. Some data have been collected indicating that a past history of infectious mononucleosis was not increased in frequency for 484 leukemics under age 16 compared with controls [80]. Fraumeni examined the history of infectious mononucleosis in 484 leukemics, their sibs, and controls obtained by intervew from parents but did not validate positive reports by means of serology. Miller et al. found no association between the EBV infection and acute leukemia in 12 newly diagnosed cases of ALL [81]. However, these studies are probably not sufficient to completely rule out an association. Even a study examining the frequency of positive serology in leukemia cases and controls could be difficult to interpret, because the onset of a clinical (or subclinical) infectious mononucleosis infection cannot with 100% certainty be determined on the basis of serology. This particular viral disorder can range from a silent or mild, subclinical infection [82] to a rather severely prolonged illness. Correlation of serology with clinical history will vary depending on the severity of clinical illness, whether physician and/or laboratory diagnosis was made, recall, etc. The most useful study design for evaluating associations between prior infection with infectious mononucleosis and subsequent development of leukemia would be a prospective approach. Follow-up should include serial serosurveys of controls to document interval occurrence or absence of infection.

Another indirect methodologic approach for examining the role of viruses in leukemogenesis has been to follow-up offspring of mothers who developed viral infections in pregnancy to determine whether these children have an increased incidence of leukemia. Similar to the situation with young kittens who are not able to mount an appropriate immunologic defense against FeLV (and therefore develop feline leukemia), it is hypothesized that the human fetus may be particularly susceptible to viral infections during pregnancy that may cause leukemia after a latent period of 3–5 years (accounting for the childhood peak in incidence).

Investigators have employed three methodological approaches to study the association between viral infections in pregnant mothers and subsequent leukemia in offspring. Most studies have followed up offspring of mothers with viral infections during pregnancy and mothers without such infections, comparing the incidence of leukemia in the two groups of offspring. Another study design used is a case-control interview, comparing leukemics to controls for history of viral illness in mothers during pregnancy. Other researchers have compared time trends for various viral diseases in a population of pregnant women with those for leukemia in children, allowing for an appropriate latent period. Key studies using these approaches are described in the subsequent section entitled, "Seasonal Incidence and Season of Birth." As noted in that section, virtually all of the studies lack serologic documentation of information about infection obtained in interviews or medical charts. Therefore, any suggestive or significant associations identified between viral infections in pregnancy and leukemia in offspring must be interpreted cautiously. Although there are a number of negative studies, the association does appear suggestive.

In 1980, Poiesz et al., working out of Gallo's laboratory at the National Cancer Institute, reported the isolation of a type-C retrovirus from fresh peripheral blood lymphocytes and from two T-cell lymphoblastoid cell lines of a black, West Indian patient with cutaneous T-cell lymphoma [4]. The virus was shown to be integrated into the genome of diseased cells, but was absent from normal cells suggesting that the virus is not genetically inherited. Screening of a number of patients with cutaneous T-cell lymphoma and Sezary T-cell leukemia has shown that most cases positive for human T-cell leukemia/lymphoma virus (HTLV) had similar clinical features including adult onset with a rapid disease course, hepatosplenomegaly, lymphadenopathy, hypercalcemia, skin manifestations, and large, pleomorphic lymphocytes with mature T-cell surface phenotype markers. The retrovirus, called HTLV, appears to be an acquired, rather than an endogenous virus and is distantly related to BLV as shown by similarity in the amino acid sequences of their major core proteins [5–9, 83]. Most American and European adult lymphoid malignancies are B-cell lymphoproliferative disorders whereas T-cell malignancies are the predominant lymphoid cancer in Japan [84]. A T-cell leukemia occurring in adults and endemic to certain areas in southern Japan, particularly Kyushu, has clinical symptoms very similar to those found in HTLV-positive patients with cutaneous T-cell lymphoma. This malignancy designated as ATL was first described by Takatsuki et al. [85–87]. More recently, patients with this disease have been found to have antibodies to HTLV as have some normal persons living in the endemic Kyushu region [88, 89]. This finding also supports horizontal transmission of HTLV. Epidemiologic studies of the HTLV-positive lymphoproliferative disorders are currently underway. Although the association between HTLV and malignant T-cell leukemias and lymphomas is supported by a number of studies, more rigorous epidemiologic methodology would further strengthen the apparent association. Such epidemiologic studies combined with appropriate viral, immunologic, and molecular approaches should delineate mechanisms and perhaps time of onset of infection, transmission, as well as sources and reservoirs of the virus. Antibodies reactive against an internal structural protein of HTLV have been identified in almost all tested cases of T-lymphosarcoma-cell leukemia born in the West Indies as well as most of the tested adult Japanese T-cell leukemia/lymphoma cases [89–91]. This suggests that the West-Indian-born patients with T-lymphosarcoma-cell leukemia, virtually all of whom were black, and the Japanese patients with adult T-cell leukemia have similar, if not identical, disorders. Additional studies are underway in both the West Indies and Japan to characterize the occurrence of the risk factors for these T-cell malignancies, which appear to be related to Sezary's syndrome and T-chronic lymphocytic leukemia but are rare in Europe and the United States.

LEUKEMIA CLUSTERS

During the past twenty years, many reports of possible clustering of leukemia cases over limited periods of time in well-defined geographic areas have been published. Such reports have been motivated by an interest in etiological factors, including transmission of viruses or other infectious organisms, time- and space-localized environmental exposures to radiation, chemicals, or animals. A critical review and analysis of the significance of various leukemia clusters reported through 1968 can be found in the review by Kessler and Lilienfeld [92]. These authors note many methodologic problems related to interpretation of the significance of such clusters including: the small number of cases in each report, the *a posteriori* character of the time and space boundaries which are drawn, the differential quality of case ascertainment in the areas of the suspected clusters in contrast to that carried out in noninvolved areas (often utilizing routine collection of mortality and incidence data for the leukemias), the differences in the various statistical methods employed to analyze such cases, and the assumptions (not always valid) used in the statistical analyses of clusters.

Knox has proposed that examination of space–time clusters for rare events requires separate examination for three components of epidemicity: concentrations in space over the entire time period of the study, concentrations in time over the whole geographic area included in the study, and interactions between space and time concentrations. In order to determine the third component, he suggests that investigators must search for high concentration areas and use a method consisting of examination of all possible pairs to analyze if short geographical distances are positively correlated with short time intervals [93]. However, variation and arbitrariness do exist in the various definitions of "close" in time and "close" in distance employed by a number of investigators, all of whom utilized the same method developed by Knox to look for evidence of leukemia clustering in different populations [94–96].

Another significant issue relates to the problem of grouping together several types of leukemias. While the Gross leukemia virus can produce, in the same inbred colony of mice, diseases ranging from ALL to histiocytic lymphoma, and the feline leukemia virus can cause lymphosarcoma, refractory anemia, or infectious peritonitis in cats, the observation that a single type of virus can produce a spectrum of diseases following a prolonged incubation period should not be generalized to man. As pointed out previously, much evidence from human data suggests that there are different risk factors for each of the different types of leukemia. It might be methodologically more sound to consider each major type of leukemia as a separate disease when identifying and analyzing potential clusters.

Clustering studies have been concerned with various types of proximity in addition to geographic and time boundaries, including school clusters (consisting of students attending a given school concurrently, or consecutively during a relatively short time period, who later developed leukemia), house clusters involving members of the same family living in a given house or people who have lived consecutively in the same house, and acquaintanceship clusters involving individuals living in a community who were directly or indirectly linked by acquaintanceship.

Several statistical techniques for analyzing potential clusters were developed during the past 15 years. In spite of the sophistication of many of these techniques, the major methodologic problems summarized above as well as variable findings (with respect to numerous "clusters" that have been subjected to rigorous scrutiny) have thrown the topic of leukemia clusters into some disfavor. However, the methodologically sound analyses that have shown a significant excess of cases in certain space–time bounded areas cannot be entirely ignored. Therefore, the statistical techniques developed to analyze clusters will be briefly reviewed. Then some of the cluster studies summarized in Tables 7-2 and 7-3 will be discussed.

In 1912–13, Pearson examined the statistical likelihood of multiple cases of disease appearing in the same house [97, 98]. Now a general approach is employed by a number of statisticians, with several variations, which involves establishment of a predetermined, somewhat arbitrary def-

Table 7-2 Analytic studies of leukemia clusters showing positive findings

Author, year	No. of cases examined	Geographic characteristics	Time period	Cell types involved	Definition of cluster	Findings
Pinkel and Nefzger, 1959 [118]	95 cases identified from local cancer registries and the New York State Cancer Registry	Urban Buffalo	1943–56	Not specified	2 or more reported cases within two years and 1/3 mile of each other	16 clusters were found ($p = 0.15$) which was suggestive of a clustering effect in Central Buffalo. Childhood leukemia had a greater incidence in the upper economic half of the population and occurred at a later average age. This investigation assumed that there was a uniform population density throughout Buffalo. Ederer et al. [99] noted that an erroneous statistical method was used. The formula for the probability of clusters was based on a fixed number of defined temporal–spatial units but the analysis involved a continuum of all possible cells of the right spatial and temporal dimensions.
Pinkel et al., 1963 [119]	95 cases identified from local and state cancer registries	Urban Buffalo	1943–56	Not specified	Aggregation is suggested by an average ridit < 0.5 for the time–space distribution of paired cases less than 1 mile apart. (0.50 is the average ridit for	The same cases and population described above were reanalyzed using a different method. The average ridits for pairs formed by leukemia cases were significantly less than 0.05

					the baseline distribution of pairs greater than 1 mile apart).	when the pairs were less than ⅛ mile apart.
Meighan and Knox, 1965 [94]	69 cases of childhood leukemia identified from death certificate review and Univ. of Oregon Medical School Hospital	City of Portland, Oregon, and its immediate environs	1950–61	Not specified	By method of Knox, 1964 [93]	The observed figures for short distances and times is consistently above those expected but never to an unequivocal degree. There are two points where the excess was found to be statistically significant: for paired cases closer than 1 km and 365 days apart and for paired cases closer than 250 days and 4 km apart.
Mustacchi, 1965 [120]	639 cases of leukemia in children and adults identified through review of medical records at every San Francisco hospital, the Univ. of Calif. S.F. hospitals, and Stanford Univ. hospitals	San Francisco	1946–55	Not specified	The 10-year study interval was divided into an initial baseline 3-year interval and a 7-year follow-up period. City census tracts were defined as "with" or "without" leukemia in the baseline period. The "observed" number of leukemia cases in the follow-up period was compared with the "expected" number on the basis of incidence in the baseline period for city tracts in which at least 1 case had occurred.	Leukemia cases in the follow-up period tended to cluster in those census tracts where cases had occurred during the baseline period. The tracts "with leukemia" in the baseline period were compared to the tracts "without" with respect to racial distribution (no difference), ethnic differences (same proportion of Jews in both groups) and crowdedness (the areas "without leukemia" had an excess of persons living alone or with one other person compared to those areas "with leukemia").

Table 7-2 (continued)

Author, year	No. of cases examined	Geographic characteristics	Time period	Cell types involved	Definition of cluster	Findings
Lundin et al., 1966 [121]	1. 130 pairs (living on same street) of leukemia–lymphoma deaths compared with 127 pairs in control group. 2. 350 subsequent deaths of leukemia–lymphoma paired with 258 index cases. Pairs came from total case population of 917 deaths from leukemia–lymphoma in Pittsburgh residents and 1273 deaths in other parts of Allegheny County	Pittsburgh, and Allegheny County, PA	1953–62	Leukemia and lymphoma, not further specified in the analysis	Excessive number of observed paired leukemia deaths during a given time interval compared with the expected number. Expected number is based on an assumption of constant probability of subsequent death in the interval between index death and the end of the study period. This assumption was tested by analyzing the mortality pattern of paired control deaths and was substantiated.	No clustering over the entire time interval. But there was an unusual temporal distribution of leukemia—an excessive number of deaths occurred between 6 and 8 years apart.
Fraumeni et al., 1966 [122]	1. 616 cases of childhood leukemia under age 5 born from 1943 through 1957 2. 1640 cases of leukemia under age 15 reported to the New York State Cancer Registry 1943–1962	New York State (excluding New York City)	1943–62	Not specified	Method of Ederer et al., 1964 [100]	No clustering by year of birth for children under 5. Examination of the occurrence of clusters according to each of four 5-year intervals of report showed the observed numbers of cases exceeding the expected for 1-year clusters in all time intervals. Investigators state that the diffuse clustering tendency in the year of report data should be interpreted cautiously because reports to

						the State were frequently delayed.
Mainwaring, 1966 [95]	74 cases of leukemia in children under 15 identified through review of medical charts at the two children's hospitals in Liverpool and from death certificates	Liverpool, England	1955–64	ALL and acute blastic leukemia AML AMoL Acute erythroleukemia	Method of Knox 1964 [93]	Excessive clustering found especially for pairs closer than 300 days and 4 km apart.
Till et al., 1967 [123]	483 children who died of leukemia under age 15 identified from death certificates	Greater London	1952–61	ALL AML AMoL Other acute "unclassified" CML	Method of Knox, 1964 [93]	Weak evidence of clustering by date and place of onset was found for all cases of ALL. Weak evidence of clustering by date, place of onset, date of birth, and place of parent's residence at birth of child for cases of ALL dying before age 6.
Gunz and Spears, 1968 [96]	1003 adults and children	New Zealand	1954–64	Children: all types of leukemia including 8 with CML Adults: acute leukemia	Method of Knox, 1964 [93]	Overall no clustering was found except in children under age 6. Four pairs developed leukemia within 2 months (3 of these within 1 month) while living distances of less than 1 mile apart.
Klauber and Mustacchi, 1970 [124]	149 leukemia cases in children under age 15 diagnosed in a 20-year time period	San Francisco	1946–65	Not specified	Method of Mantel, 1967 [103]	A comparison of "within" time interval average distance between cases to "between" time interval average distance indicated that clustering in the series was weak. The clustering appeared to be the result of a slight excess of pairs within relatively large time and/or space distances.

Table 7-2 (continued)

Author, year	No. of cases examined	Geographic characteristics	Time period	Cell types involved	Definition of cluster	Findings
Zahalkova et al., 1970 [125]	104 cases 0–15 years of age at the time of diagnosis	13 districts in Southern Moravia, a province in Czechoslovakia	1960–68	Not specified	Method of Knox [93] and Pinkel and Nefzger [118]	Various space and time intervals were evaluated. A significant excess of observed compared with expected number of pairs was found for distances of 0–12 km and both 0–1 year time intervals and 0–2 year intervals. The authors noted that the incidence of acute leukemia did not differ remarkably from that of other countries.
Larsen et al., 1973 [126]	595 cases of acute leukemia	Metropolitan Atlanta	1956–69	ALL ANLL	A combinatorial test for measuring unimodal clustering was developed and tested by these authors. It is related to the David and Barton method but differs in that the "center" of the Larsen et al. cluster is determined by the locations of the successes and not in an *a priori* fashion.	A tendency of ALL cases ages 0–15 was noted to cluster. To a lesser extent cases of ANLL ages 50+ were also noted to cluster. The authors note that the total population in Metropolitan Atlanta increased 37% during 1960–70 but the age distribution in the various counties remained fairly stable. However, incidence of both types of leukemia apparently remained stable and there was not a tendency for cases to occur in later years.

Smith et al., 1976 [127]	623 children dying of ALL under age 6	Greater London	1952–65	ALL	Method of Pike and Smith, 1974 [107]	Slight evidence of clustering was found associated with cases with onset 1952–60 but no evidence for cases with onset 1961–64.
Greenwald et al., 1979 [106]	37 cases first diagnosed as having leukemia or lymphoma 1967–72 (out of 42 who met the criteria, 5 refused interview)	Orleans County, New York State	1967–72	14 lymphosarcoma and reticulum cell sarcoma 5 Hodgkin's disease 1 other lymphoma 10 lymphatic leukemia 5 myeloid leukemia 1 monocytic leukemia 1 unspecified leukemia	Based on methods of Mantel; (1967, 1970) [103,109,110] 1. Comparison of acquaintance-linked cases with acquaintance-linked controls. 2. Computer simulation of "disease transmission". 3. Calculation of secondary attack rates. See text for description of these methods.	1. Comparison of linkage of case pairs with that of control pairs showed statistically significant case-control differences. 2. Computer simulation of disease transmission from selected source cases to selected targets and controls showed the transmissability hypothesis to be well supported for the lymphomas and the leukemias. 3. Secondary attack rates were higher for cases than for different sets of matched controls but these results were not subjected to quantitative hypothesis testing. Conclusion: With two intermediate acquaintances there were significant differences in the amount of linkage between cases compared with that between matched controls. The results are consistent with a transmissability hypothesis, possibly involving transmission via asymptomatic intermediaries.

Table 7-3 Analytic studies of leukemia clusters showing negative findings

Author, year	No. of cases examined	Geographic characteristics	Time period	Cell types involved	Definition of cluster	Findings
Ederer et al., 1964 [99] Ederer et al., 1965 [100]	2610 cases of leukemia in children and adults identified from the Connecticut State Cancer Registry. 333 children under 15 years of age	State of Connecticut	1945–59	Not specified	Method of Ederer et al., 1964 [99]	No evidence of clustering in either 1- or 2-year clusters for children under either 15 or of all ages. Clustering was found for both polio and hepatitis using the methodology.
Barton et al., 1965 [101]	84 cases of acute leukemia all ages	Liverpool, England	1957–62	Acute leukemia, not otherwise specified	Method of Barton et al., 1965 [101] See text	No evidence for clustering for any of these case groups with leukemia. However, the method did show evidence of significant clustering for measles and polio.
	54 cases of acute leukemia all ages	Bristol, England	1956–62	Acute leukemia, not otherwise specified		
	118 cases of acute leukemia all ages	Birmingham, England	1954–61	Acute leukemia, not otherwise specified		
	73 cases of acute leukemia all ages	Sheffield, England	1954–61	Acute leukemia, not otherwise specified		
	96 children under 6 years of age. This case group was analyzed earlier by Knox, 1964 [93]	Northumberland and Durham, England	1951–60	Acute leukemia in children under 6 years. Not otherwise specified		
	85 cases of acute leukemia	Cornwall, England	1947–62	Acute leukemia, not otherwise specified		

Stark and Mantel, 1967 [128]	706 cases of leukemia of all ages identified by death certificate	State of Michigan. Data also examined county by county	1950–64	Not specified	Method of Mantel, 1967 [103]	No evidence of clustering when analyzed: (a) by the entire State of Michigan (b) county by county
Klauber, 1968 [129]	234 children who died of leukemia before age 5	State of California	1958–60	Not specified	Method of Mantel, 1967 [103]	No significant clustering by hospital of birth. No significant within hospital clustering by day of birth for leukemia cases
Glass and Mantel, 1969 [130]	298 white children dying of leukemia before their 15th birthday	Los Angeles County, CA	1960–64	Not specified	Method of Knox, 1964 [93], and Mantel, 1967 [103]	No evidence of time–space clustering when deaths examined by age at death in three groupings: 0–14, 0–5, 2–9
Bailar et al., 1970 [131]	4552 all ages reported to the Cancer Registry (including 794 reported only by death certificate)	169 towns in Connecticut	1935–63	Not specified	The number of pairs of cases occurring at fixed time intervals in the same small areas is used as a measure of epidemicity.	No evidence of time–space clustering. A barely significant deficit of cases at 6-year intervals was found for all ages. No evidence of clustering when the data was stratified by age (0–14, 15+) or by town (towns with fewer than 10 cases, towns with 10–19 cases, towns with 20 or more cases).

Table 7-3 (continued)

Author, year	No. of cases examined	Geographic characteristics	Time period	Cell types involved	Definition of cluster	Findings
Glass et al., 1971 [132]	272 cases of acute leukemia ages 0–15 at time of death	New Zealand	1953–64	Acute childhood leukemia, not otherwise specified	Method of Knox, 1964 [93], and Mantel, 1967 [103]	Earlier analysis of data showed statistically significant time–space clustering in children less than 6 years old at diagnosis but not in children older than 6 (Gunz and Spears). When analysis was confined to successive 6-year periods or an intermediate 6-year period rather than the entire 12-year period, evidence of clustering in children under 6 diminished. Authors suggest that the marked growth of population in New Zealand in this time interval and major population shifts from urban and suburban areas not taken into account in the initial analysis may have resulted in a spurious appearance of clustering.
Alperovitch et al., 1974 [133]	983 cases of all leukemias and lymphomas in all age groups	7 rural French regions	1967–69	ALL AML CLL CML Lymphosarcomas and reticulum cell sarcomas	Methods of Knox, 1964 [93], and Mantel, 1967 [103]	No clustering tendency in any region for any cell type of leukemia.

Zack et al., 1977 [134]	500 cases of leukemias, lymphomas, and Hodgkin's disease diagnosed between 15 and 29 years of age of 604 eligible cases identified from the Connecticut Tumor Registry	State of Connecticut	1960–71	Leukemias not further specified Lymphomas not further specified Hodgkin's disease	1. Case-control approach comparing the number of patients simultaneously attending the same high school with the number predicted from simulated controls. 2. Cohort approach comparing disease rates in persons enrolled at the same high schools at the same time as index leukemia, lymphoma, HD patients (exposed cohort) compared with rates in persons enrolled in other high schools (nonexposed cohort). 3. Cohort approach in which the observation period was divided into halves. High schools with and without cases in the first half were compared with respect to the number of cases in each high school during the second half.	1. No significant increase in intradisease case–case links was found for leukemia (statistically significant increases were found for Hodgkin's disease). 2. Leukemia was not diagnosed more often in the exposed than in the unexposed. (HD was found more often in exposed compared with nonexposed.) 3. Cases diagnosed in the first period were not associated with a markedly larger number of cases in the second period in the same school compared with a control school.

inition of units of time and space for comparing observed numbers of leukemia cases to expected numbers based on incidence or mortality data or other measures. Ederer et al. pointed out that it is important to define units of space and time before examining the data in detail otherwise the investigator will be tempted to gerrymander [99, 100]. These investigators noted that the size of the units is very important; clusters may not be detected if the unit of study is too large or too small.

Other examples employing defined units of space and time include methods proposed by Knox [93] and David [101, 102]. Knox used a 2 × 2 contingency table approach for the set of all spatial (residence) and temporal (time of disease onset) differences to determine whether there were more leukemia cases living close together and developing the disease around the same time than one would have expected from the general distribution of patients by space and time [93]. Determination of statistical significance is based on an *a priori* definition of what is meant by "close in time" and "close in space." Barton and David suggested a method for determining the null distribution of the number of close pairs and its variance by considering two maps: a time map and a space map [101]. The intersection of the two maps in the number of close pairs both in time and in space. They also developed a procedure for determining the exact permutational variance for Knox's number-of-close-pairs statistic [102], making dependence upon the Poisson assumption unnecessary.

After critically evaluating the methods described by Knox and Barton and David, Mantel devised a more generalized regression approach that considers both the Knox and Barton-David procedures as special cases [103]. Klauber extended Mantel's approach for testing space–time clustering of a single set of points to be able to test for clustering between two such sets of points [104]. He proposed an empirical randomization test together with its normal approximation for two situations: (a) one set of points is considered fixed and the other random, and (b) both sets of points are considered random. For either situation, a variety of space and time distance measures may be used.

Two additional methodologic strategies focusing on contact between pairs of cases have been proposed. Pike and Smith closely examined the period of common exposure for pairs of cases [105] while Greenwald, Rose, and Daitch analyzed closeness of acquaintance linkage between matched pair of cases compared with pairs of noncases [106].

Pike and Smith postulated two time periods, a period of susceptibility, during which the case "caught" the leukemia, and the period of infectivity, during which the case was "infective" [105]. They also postulated two areas of space representing the patient's "effective" movements during his period of susceptibility and period of infectivity, respectively. The model postulated is that of pairs of cases, one of whom was in the appropriate time–space "location" to catch the disease (leukemia) from the other person. In order to determine whether a significant increase in such pairs has occurred compared with an expected number, it is necessary to know the expected distribution of such pairs. Pike and Smith derived expressions for the expected distribution of X such pairs and the variance of X. Their method attempts to provide a conceptually appropriate technique for examining potential clusters of a disease which has a long latent period and for which associated cases may have had dates of onset several years apart. This represents an extension of Knox's method [93].

In 1974, Pike and Smith described a case-control approach to examine diseases (including those with long latent periods) for evidence of contagion. Using this approach, one can evaluate the statistical significance of the total "effective" contact between all possible pairs of patients [107]. In addition, one can determine whether the number of patients involved in "effective" contact with at least one other patient is significantly greater than expected. These authors noted that selection of a satisfactory control group is crucial and discuss necessary matching factors in a later paper [108].

Greenwald, Rose, and Daitch [106] developed a method for comparing linkage of case pairs with that of control pairs based on the methods of Mantel and co-workers [103, 109, 110]. Instead of using space and time as variables, these investigators used closeness of acquaintance linkage and categorization as "case–case" or "*not* case–case." In addition, they carried out a computer simulation of the transmission of the hypothetical diseases among the actual study subjects during the study period. To carry out this simulation, Greenwald et al. developed a probability model of disease transmission which defined a summary measure of acquaintance, sometimes involving intermediaries, based on the probability of transmission of a hypothetical disease from one individual to another. Finally, a secondary attack rate approach was used. These investigators defined a "secondary attack rate" for a group of cases as the number of cases linked to at least one other case divided by the total number of distinct individuals linked to the group of cases. For a group of controls, the "secondary attack rate" is defined as the number of cases linked to at least one control divided by the number of distinct individuals linked to the group of controls. (As pointed out by the investigators, an ordinary secondary attack rate cannot be calculated because of the lack of identification of an index case.)

In addition to differences in statistical methods and tests for evaluating potential clusters, other methodologic problems must be taken into account. Even if some proportion of leukemia cases are caused by a transmissable agent, members of pairs of cases in contact with each other may have actually had a common exposure to such an agent at some point in time, rather than one of the cases transmitting the putative leukemogenic virus to another.

Numerous leukemia clusters have been reported. One of the earliest of these was a report of eight acute childhood leukemia cases occurring in Niles, Illinois, in the four-year period of 1957–60 [111]. Seven of the eight children were noted to have contacts in one particular school in the community. An occurrence of a "rheumaticlike" illness that involved a number of children at the same school during the same time period was reported. The authors also found that seven infant deaths from congenital heart lesions had occurred from 1955 to 1960. This detailed report was followed by another by Heath, Manning, and Zelkowitz (involving apparent clusters of acute childhood leukemia) [112] as well as a number described in the Leukemia Surveillance Program Report, periodically issued for a number of years by the Centers for Disease Control (including some consisting of acute childhood leukemia cases only, as well as others consisting of both acute and chronic leukemia), and more recent reports by Gilmore and Zelesnick [113], McPhedran and Heath [114], Kemmoona [115], Bartsch [116], and Reid [117].

The main analytic studies of leukemia clusters are summarized in Tables 7-2 and 7-3, grouped into those in which clustering was found and those that did not show good evidence for clustering. Leukemia cell type is infrequently specified in the majority of analytic studies. Although a few of the analytic studies focus on leukemia cases from one or two hospitals, the majority consist of broader population-based identification of most or all cases from hospital records, hospital and state tumor registries, and death certificates. In spite of differences between the studies in definitions of time and space units used, statistical methods of analysis applied, age groups studied, histopathologic classifications or groupings employed, cities, counties, states, or countries included, outcome measures utilized (i.e., date of diagnosis of leukemia versus date of reporting to a registry versus date of death), and occasional failure to include in the analysis secular changes in denominators over the time period being studied, several of the studies consistently demonstrated an apparent clustering effect, albeit weak in many cases, for childhood leukemia. In the positive studies showing evidence of clustering for childhood leukemia where histopathologic types were separately analyzed, the positive findings or significant excess of cases in certain time–space units appear to be confined to cases of ALL, particularly children under age 6. There are also a number of negative studies of clustering in this age group for children with acute leukemia; however, it should be pointed out that virtually all of these negative studies used cases ascertained from death certificates whereas significant proportions of the positive studies used incident cases obtained from medical records and tumor registries.

SEASONAL INCIDENCE AND SEASON OF BIRTH

Many infectious and some chronic diseases have characteristic seasonal onsets and/or exacerbations related to seasonal occurrence of etiologic or precipitating factors. A number of investigators have therefore examined the seasonal distribution of leukemia cases. Such an association would support the concept that a subgroup of one or more types of leukemia had an epidemic basis. Many difficult methodologic questions must be answered before such apparent seasonal associations can be accepted as real rather than artifactual. The leukemias often have somewhat insidious onsets and the date of onset may be hard to pinpoint. Various investigators have defined it differently such as onset of symptoms, onset of clinical signs, date of hospital admission during which the diagnosis was made, date of bone marrow examination that established the diagnosis, date of discharge from the diagnostic admission. Because the leukemias are rare, results in most studies of seasonality are based on small numbers of cases. As in many other epidemiologic investigations, the case group is often a select series derived from admission or clinic visits to one or more hospitals or death certificate series, rather than a population-based identification of all incident cases. In addition, seasonal associations which seem to be related to etiology may actually reflect seasonal occurrence of respiratory or other infections to which persons with diagnosed and undiagnosed leukemia are actually more susceptible. Also, many studies are not histopathologic-type specific.

In view of these many complex methodologic issues, it is difficult to know how to interpret the variation in findings. Guasch analyzed season of onset for 200 leukemics, all histopathologic types and all ages, seen at a single clinic in Barcelona during the 1940's and 1950's. Onset of initial clinical signs appeared to show a winter peak [135]. Hayes found peak onset of occurrence between December and April for 184 persons of all ages admitted to a single U.S. hospital in North Carolina, 1943–58. No apparent association was found between leukemia onset and seasonal variation for measles, pertussis, scarlet fever, and respiratory infections [136]. Fraumeni also found an excess of leukemia onsets in winter and spring for 511 cases under age 16 seen at 12 medical centers in the U.S. [137]. Swan noted winter and spring peaks of disease onset (as well as date of death) for 56 adults with acute leukemia admitted to one London hospital [138].

Several studies noted a summer peak for onset of leukemia. Lee examined season of onset for 2,103 leukemics under age 20 and 2,570 leukemics ages 20–44 reported to the National Cancer Registration Scheme in England and Wales 1946–60. He noted two peaks, one in May–June and another in December–January, which were more prominent for those with lymphatic leukemia and which also increased with age [139, 140]. A second similar study carried out by Lee in New Zealand, 1958–61, and Australia, 1946–59, of a total of 1,530 leukemics with all cell types and all ages also showed both a summer peak and a November–December peak of onset [141]. Knox found a summer peak especially for those with lymphoblastic leukemia and in children under 6 in a study of 185 children under age 15 identified from hospitals, cancer registries, and death certificates 1951–60 in Northumberland and Durham, England [94]. Mainwaring also found a summer peak of onset for children with lymphatic leukemia and those under 5 in a study of 74 leukemics under age 15 seen in hospitals in Liverpool and identified from death certificates, 1955–65 [142]. Von Ambs et al. noted a June peak in clinical onset for 85 children under age 15 diagnosed as having leukemia at a German pediatric clinic, 1947–64 [143].

Other studies have shown no seasonal peak for onset of all leukemias and specifically lymphocytic leukemias [94, 96, 133, 144]. A large recent analysis of dates of diagnosis for detection of seasonal occurrence of 7,000 cases of acute leukemia, 1969–81, showed no evidence of seasonality. The cases were identified from the Third National Cancer Survey (1969–71), the Surveillance, Epidemiology, and End Results (SEER) Program (1973–77), and the Centralized Cancer Patient Data System (CCPDS) with data from 21 cancer centers (1977–81). Further examination of the data by histopathologic type, age, or geographic location also showed no evidence of seasonal excess [145]. In view of the variability of findings as well as the lack of association noted in a number of studies, it is difficult to ascribe any significance to peaks in seasonal occurrence observed in some studies [146].

Birthdates and season of birth of leukemic children have been examined as another methodologic approach used to determine whether periodically or seasonally occurring viral infections might be leukemogenic. Most of the investigations studying the postulated association between dates of birth and epidemic infections have not documented the occurrence of infection prior to development of leukemia by serologic testing in leukemia cases or in their mothers.

Several population-based studies have shown no excess of leukemic births in any particular season or years [122, 128, 146, 147]. The number of cases included in each of the studies ranged from 400 to 2,000 leukemic children. But Ederer found an excess of May–June births and a deficit of October–November births among 961 white infants dying of leukemia,

1950–59, in the U.S., when compared with total U.S. births and births of all infants dying of all other neoplasms during 1950–59 [148]. The focus on season of birth seems rather arbitrary when a more appropriate relationship to investigate is probably the association between the occurrence of epidemics of various viral disorders, particularly in pregnant women, and subsequent incidence of leukemia in their offspring.

Several studies have examined associations between epidemics of influenza and chicken pox affecting pregnant women and occurrence of leukemia in their children. The general approach has been to follow up infants and children of substantial numbers of affected pregnant women and compare leukemia incidence to that occurring in offspring of women unaffected by such viral diseases during their pregnancies. A follow-up study of 1,959 infants born in the first week of March 1958 to mothers reported to have influenza during pregnancy showed a fivefold excess risk of hematopoietic malignancies including leukemia (mostly ALL), lymphoma, and Hodgkin's disease compared with 14,791 infants of mothers who had not had influenza [149]. The same authors also found a correlation between two peaks in the occurrence of influenza during 1955–64 (as measured by sickness–absence statistics) and subsequent peaks in leukemia deaths in children under 5 whose birthdates were consistent with pregnancy-related exposure to the influenza epidemics. Another study demonstrating a possible link between influenza in pregnant women and subsequent leukemia in their offspring was carried out in Finland by Hakulinen et al. These investigators compared birthdates of leukemics, identified from a cancer registry, to dates of epidemics of influenza, identified from sickness–absence statistics and routine laboratory reports of viral studies. They found an increased incidence of leukemia in children born shortly after the 1957 epidemic but no increase in children born after other epidemics [150].

Evidence for an association between varicella and childhood acute leukemia comes from a study by Vianna and Polan comparing incidence trends for various viral diseases and mortality trends for acute leukemia from 1936 through 1963. Of the various viral diseases examined, only varicella incidence was found to parallel acute leukemia mortality, allowing for a latency period. A retrospective search of varicella case records identified 63 instances of varicella complicating pregnancy. Three children resulting from these 63 pregnancies subsequently developed ALL [151]. A follow-up of 1,016 children born in 1950–52 to mothers who had chicken pox, mumps, or rubella during pregnancy showed two deaths from leukemia (both born to mothers who had chicken pox), which was a significant excess compared to expected [152].

There have also been a few negative reports showing no association between epidemics of viral disorders, such as influenza and chicken pox, particularly affecting pregnant women and subsequent increase in leukemia in their offspring. Randolph and Heath analyzed the incidence of leukemia among cohorts of children less than 15 years of age living in Atlanta and Houston in relationship to birthdates occurring during the period 1957–71. The investigators defined births as influenza or non-influenza associated depending on whether the birth occurred during a period of influenza/pneumonia mortality excess. Using four age groupings and five diagnostic classifications, rates of leukemia were compared in children who were defined as exposed in utero to those not exposed. No clear-cut asociations between exposure to influenza and increase in leukemia incidence for any of the age or diagnostic categories was found [153]. Leck and Stewart examined leukemia death rates for children born a few months after six influenza epidemics in the period 1951–68. No apparent leukemia excess was observed [154].

The evidence is somewhat conflicting but, in general, it appears that infections in pregnant women, particularly varicella and influenza, may increase subsequent risk of leukemia in children. Additional studies employing serologic confirmation, immunologic studies in pregnant women who develop viral infections and their offspring, as well as analysis by leukemia cell type appear warranted.

REFERENCES

1. V. Ellerman and O. Bang, "Experimentalle leukämie bei huhnern," *Centr. Bakteriol. Abt. 1 Orig.* **46:**595–609 (1908).
2. L. Gross, "Viral etiology of cancer and leukemia: a look into the past, present, and future—G.H.A. Clowes Memorial Lecture," *Cancer Res.* **38:**485–493 (1978).
3. R. Lewin, "New reports of a human leukemia virus," *Science* **214:**530–531 (1981).
4. B.J. Poiesz, F.W. Ruscetti, A.F. Gazdar, P.A. Bunn, J.D. Minna, and R.C. Gallo, "Detection and isolation of type C retrovirus particles from fresh and cultured lymphocytes of a patient with cutaneous T-cell lymphoma," *Proc. Natl. Acad. Sci. USA* **77:**7415–7419 (1980).
5. Y. Hinuma, K. Nagata, M. Hanaoka, M. Nakai, T. Matsumoto, K. Kinoshito, S. Shirakawa, and I. Miyoshi, "Adult T-cell leukemia: antigen in an ATL cell line and detection of antibodies to the antigen in human sera," *Proc. Natl. Acad. Sci. USA* **78:**6476–6480 (1981).
6. I. Miyoshi, I. Kubonishi, S. Yoshimoto, I. Akagi, Y. Ohtsuki, Y. Shiraishi, K. Nagata, and V. Hinuma, "Type C virus particles in a cord T-cell line derived by co-cultivating normal human leukocytes and human leukaemic T-cells," *Nature* **294:**770–771 (1981).

7. D. Catovsky, M.F. Greaves, M. Rose, D. Galton, A. Goolden, D. McCluskey, J. White, I. Lampert, G. Bourikus, R. Ireland, A.I. Brownell, J. Bridges, W. Blattner, and R. Gallo, "Adult T-cell lymphoma-leukemia in blacks from the West Indies," *Lancet* **1:**639–643 (1982).
8. R.C. Gallo and M.S. Reitz, "Human retroviruses and adult T-cell leukemia-lymphoma," *JNCI* **69:**1209–1212 (1982).
9. M. Essex, "Meeting highlights. Adult T-cell leukemia-lymphoma: role of a human retrovirus," *JNCI* **69:**981–985 (1982).
10. L. Tyler, "Enzootic bovine leucosis," *Vet. Rec.* **103:**194–198 (1978).
11. C. Olson, "Bovine lymphosarcoma (leukemia)—a synopsis," *J. Am. Vet. Med. Assoc.* **165:**630–632 (1974).
12. O.C. Straub and E. Weinhold, "Zur frage der übertragung boviner leukose durch Kolostrum and Milch," *Dtsch. Tierarztl. Wochenschr.* **78:**441–444 (1971).
13. J.F. Ferrer and C.E. Piper, "An evaluation of the role of milk in the natural transmission of BLV," *Ann. Rech. Vet.* **9:**803–807.
14. M. Mussgay and O.R. Kaaden, "Progress in studies on the etiology and serologic diagnosis of enzootic bovine leukosis," *Curr. Top. Microbiol. Immunol.* **79:**43–72 (1978).
15. O.C. Straub, "Horizontal transmission studies on enzootic bovine leukosis," *Ann. Rech. Vet.* **9:**809–813 (1978).
16. G. Migaki, "Hematopoietic neoplasms of slaughter animals," *Natl. Cancer Inst. Monogr.* **32:**121–139 (1969).
17. N.C. Pedersen, G. Theilen, M.A. Keane, L. Fairbanks, T. Mason, B. Orser, C. Chen, and C. Allison, "Studies of naturally transmitted feline leukemia virus infection," *Am. J. Vet. Res.* **38:**1523–1531 (1977).
18. M. Essex, "Immunity to leukemia, lymphoma, and fibrosarcoma in cats: a case for immuno-surveillance," in *Contemporary Topics in Immunology,* Vol. 6, pp. 71–106. (M.G. Hanna, Jr., ed). New York: Plenum Press, 1977.
19. E.A. Hoover, R.G. Olsen, W.D. Hardy, Jr., J.P. Schaller, and L.E. Mathes, "Feline leukemia virus infection: age-related variation in response of cats to experimental infection," *J. Natl. Cancer Inst.* **57:**365–369 (1976).
20. D.P. Francis and M. Essex, "Leukemia and lymphoma: infrequent manifestations of common viral infections? A review," *J. Infect. Dis.* **138:**916–923 (1978).
21. H.G. Purchase and B.R. Burmester, "Leukosis/sarcoma group," in *Diseases of Poultry,* 6th ed, pp. 502–568 (M.S. Hofstad, B.W. Calnek, C.S. Helmboldt, W.M. Reid, and H.W. Yoder, Jr., eds). Ames, Iowa: Iowa State University Press, 1972.
22. B.R. Burmester, "The shedding of the virus of visceral lymphoma-

tosis in the saliva and the feces of 'normal' lymphomatous chickens," *Poultry Sci.* **35:**1089–1099 (1956).

23. H. Rubin, L. Fanshier, A. Cornelius, and W.F. Hughes, "Tolerance and immunity in chickens after congenital and contact infection with an avian leukosis virus," *Virology* **17:**143–156 (1962).
24. H.R. Morgan, "Antibodies for Rous sarcoma virus (Bryan) in fowl, animal and human populations of East Africa. II. Antibodies in domestic chickens, wild fowl, primates and man in Kenya, and antibodies for Burkitt lymphoma cells in man," *J. Natl. Cancer Inst.* **39:**1229–1234 (1967).
25. L.G. Wolfe, L.A. Falk, and F. Deinhardt, "Oncogenicity of herpes virus saimiri in marmoset monkeys," *J. Natl. Cancer Inst.* **47:**1145–1162 (1971).
26. G. Theilen, D. Gould, M. Fowler, and D.L. Dungworth, "C-type virus in tumor tissue of a wooley monkey (lagothrix spp) with fibrosarcoma," *J. Natl. Cancer Inst.* **47:**881–889 (1971).
27. T. Kawakami, S.D. Huff, P.M. Buckley, D.L. Dungworth, and S.P. Snyder, "C-type virus associated with gibbon lymphosarcoma," *Nat. New Biol.* **235:**170–171 (1972).
28. L.V. Meléndez, R.D. Hunt, M.D. Daniel, B.J. Blake, and F.G. Garcia, "Acute lymphocytic leukemia in owl monkeys inoculated with herpes virus saimiri," *Science* **171:**1161–1163 (1971).
29. L.A. Falk, L.G. Wolfe, and F. Deinhardt, "Oncogenesis of herpes virus saimiri in marmosets," *Fed. Proc.* (abs) **30:**514 (1971).
30. A. DePaoli, D. Johnson, and W.W. Noll, "Granulocytic leukemia in white handed gibbons," *J. Am. Vet. Med. Assoc.* **163:**624–628 (1973).
31. H.H. Hoss, and C. Olson, "Infectivity of bovine C-type virus for sheep and goats," *Am. J. Vet. Res.* **35:**633–637 (1974).
32. C. Olson, L.D. Miller, J.M. Miller, and H.E. Hoss, "Brief communication: transmission of lymphosarcoma from cattle to sheep," *J. Natl. Cancer Inst.* **49:**1463–1467 (1972).
33. H.M. McClure, M.E. Keeling, R. Custer, R.R. Marshak, D.A. Abt, and J.F. Ferrer, "Erythroleukemia in two infant chimpanzees fed milk from cows naturally infected with bovine C-type virus," *Cancer Res.* **34:**2745–2757 (1974).
34. H. Ritter, "Studien über die tragurgswege bei der enzootische rinderleukose," *Dtsch. Tierraerztl. Wochenschr.* **72:**56–60 (1965).
35. H.M. Lemon, M.J. Twiehaus, R.B. Wilson, P.G. Rigby, C.A. Mebus, and W. Nimocks, "Symbiotic human and bovine lymphoma," *Proc. Am. Assoc. Cancer Res.* (abs) **7:**41 (1966).
36. D. Wisniewski and J. Weinreich, "Lymphatische leukämie bei vater und sohn," *Blut* **12:**241–244 (1966).

37. J. Aleksandrowicz, "Leukemia in humans and animals in the light of epidemiologic studies with reference to problems of its prevention," *Acta Med. Pol.* **9:**217–230 (1968).
38. C.W. Heath, Jr., "Human leukemia: genetic and environmental clusters," *Bibl. Haematol.* **36:**649–653 (1970).
39. C. Bartsch, F. Springher, and H. Falk, "Acute nonlymphocytic leukemia. An adult cluster," *J. Am. Med. Assoc.* **232:**1333–1336 (1975).
40. S.J. Cutler and J.L. Young (eds), *Third National Cancer Survey.* National Cancer Institute Monograph No. 41. Washington, D.C.: U.S. Government Printing Office, 1975.
41. J. Waterhouse, C. Muir, P. Correa, and J. Powell (eds), *Cancer Incidence in Five Continents,* Vol III, Sci. Publ. No. 15. Lyon, France: I.A.R.C., 1976.
42. H.J. Bendixen, "The use of epizootiological experience as a guide for the study of etiologic pathogenesis and diagnosis of bovine leukosis," *Vet. Microbiol.* **1:**281–285 (1976).
43. W. Wittmann and D. Urbaneck, "Leukosen des Rindes," in *Handbuch der Virusinfektionen bei Tieren,* Bd. V/l, pp. 42–174 (H. Rohrer, ed). Jena: Gustav Fischer Verlag, 1969.
44. E. Henriksen and B.P. Jensen, "Bovine leukosis, human leukemia and related diseases," *Ugeskr. Laeg.* **133:**1201–1205 (1971).
45. R. Flamant, M. Hayat, and D. Schwartz, "Etude des leucémies humanines en relation avec les leucoses bovine," *Eur. J. Cancer* **3:**17–19 (1967).
46. N.K. Jensen, "A topographical study of leukemia in man and cattle in Denmark," *Bibl. Haematol.* **31:**326–330 (1968).
47. E. Johnson, personal communication, 1982.
48. R.K. Anderson, D.K. Sorensen, V. Perman, V.A. Dirks, M.M. Snyder, and J.E. Bearman, "Selected epizootiologic aspects of bovine leukemia in Minnesota (1961–65)," *Am. J. Vet. Res.* **32:**563–577 (1971).
49. E. Fasal, E.W. Jackson, and M.R. Klauber, "Leukemia and lymphoma mortality and farm residence," Am. J. Epidemiol. **87:**267–274(1968).
50. S. Milham, Jr., *Occupational Mortality in Washington State, 1950–71,* Vol. 1. Contract No. 99074-26. Washington, D.C.: U.S. D.H.E.W., P.H.S., C.D.C., N.I.O.S.H., 1976.
51. L. Houten, I.D.J. Bross, and E. Viadana, *A Retrospective Survey of Cancer in Relation to Occupation.* Contract No. HSM 99-73-5. Washington, D.C.: U.S.D.H.E.W., P.H.S., C.D.C., N.I.O.S.H., 1977.
52. A.M. Adelstein, "Occupational mortality: cancer," *Ann. Occup. Hyg.* **15:**53–57 (1972).

53. A. Linos, R.A. Kyle, L.R. Elveback, and L.T. Kurland, "Leukemia in Olmsted County, Minnesota, 1965–1974," *Mayo Clin. Proc.* **53:**714–718 (1978).
54. S. Milham, Jr., "Leukemia and multiple myeloma in farmers," *Am. J. Epidemiol.* **94:**307–310 (1971).
55. A. Blair and T.L. Thomas, "Leukemia among Nebraska farmers: a death certificate study," *Am. J. Epidemiol.* **110:**264–273 (1979).
56. K.J. Donham, J.W. Berg, and R.S. Sawin, "Epidemiologic relationships of the bovine population and human leukemia in Iowa," *Am. J. Epidemiol.* **112:**80–92 (1980).
57. E. Kvarnfors, B. Henricson, and G. Hugoson, " statistical study on farm and village level in the possible relations between human leukemia and bovine leukosis," *Acta Vet. Scand.* **16:**163–169 (1975).
58. I.D.J. Bross and R. Gibson, "Cats and childhood leukemia," *J. Med.* **1:**180–187 (1970).
59. I.D.J. Bross, R. Beitell, and R. Gibson, "Pets and adult leukemia," *Am. J. Pub. Health* **62:**1520–1531 (1972).
60. B. Hanes, M.B. Gardner, C.G. Loosli, G. Heidbreder, B. Kogan, H. Marylander, and R.J. Huebner, "Pet association with selected human cancer: a household questionnaire survey," *J. Natl. Cancer Inst.* **45:**1155–1162 (1970).
61. M.B. Gardner, "Current information on feline and canine cancers and relationship or lack of relationship to human cancer," *J. Natl. Cancer Inst.* **46:**281–290 (1971).
62. R.C. Schneider, "The natural history of malignant lymphoma and sarcoma in cats and their associations with cancer in man and dog," *J. Am. Vet. Med. Assoc.* **157:**1753–1758 (1970).
63. R. Schneider, "Human cancer in households containing cats with malignant lymphoma," *Int. J. Cancer* **10:**338–344 (1972).
64. S. Izui, P.H. Lambert, N. Carpentier, and P.A. Miescher, "The occurrence of antibodies against single-stranded DNA in the sera of patients with acute and chronic leukaemia," *Clin. Exp. Immunol.* **24:**379–384 (1976).
65. G.R.V. Hughes, S.A. Cohen, and C.L. Christian, "Anti-DNA activity in systemic lupus erythematosus: a diagnostic and therapeutic guide," *Ann. Rheum. Dis.* **30:**259–264(1971).
66. D. Koffler, R.I. Carr, V. Agnello, T. Fiezi, and H.G. Kunkel, "Antibodies to polynucleotides: distribution in human serums," *Science* **166:**1648–1649 (1969).
67. G.R.V. Hughes, "Frequency of anti-DNA antibodies in SLE, RA and other diseases," *Scand. J. Rheum. Suppl.* **11:**42–51 (1975).
68. H.B. Lindsley, S. Kysela, and A.D. Steinberg, "Nucleic acid anti-

bodies in African trypanosomiasis: studies in rhesus monkeys and man," *J. Immunol.* **113:**1921–1927 (1974).

69. G. Gahrton, B. Wahren, D. Killander, and G.E. Foley, "Epstein-Barr and other herpes virus antibodies in children with acute leukemia," *Int. J. Cancer* **8:**242–249 (1971).
70. P.H. Levine, D.A. Merrill, N.C. Bethlenfalvay, L. Babich, D.A. Stevens, and D.E. Waggoner, "A longitudinal comparison of antibodies to Epstein-Barr virus and clinical parameters in chronic lymphocytic leukemia and chronic myelocytic leukemia," *Blood* **38:**479–484 (1971).
71. W. Henle and G. Henle, "Epstein-Barr virus and human malignancies," *Cancer* **34:**1368–1374 (1974).
72. J. Melnick, E. Adam, and S.E. Rawls, "The causative role of herpes virus type 2 in cervical cancer," *Cancer* **34:**1375–1385 (1974).
73. W. Dameshek and F. Gunz, "The lymphoproliferative disorders," in *Leukemia,* pp. 403–423 (W. Dameshek and F. Gunz, eds). New York: Grune & Stratton, 1964.
74. B.L. Lapkin, L. Canales, and A.M. Mauer, "Infectious mononucleosis in a child with acute lymphocytic leukemia," *J. Pediatr.* **71:**876–877 (1967).
75. D.A. Stevens, P.H. Levine, S.K. Lee, M.J. Sonley and D.E. Waggoner, "Concurrent infectious mononucleosis and acute leukemia," *Am. J. Med.* **50:**208–217 (1971).
76. A.H. Ragab and T.J. Vietti, "Infectious mononucleosis, lymphoblastic leukemia and the E-B virus," *Cancer* **24:**261–265 (1969).
77. W.L. Deardorff, P. Gerber, and W.R. Vogler, "Infectious mononucleosis in acute leukemia with rising Epstein-Barr virus antibody titers," *Ann. Intern. Med.* **72:**235–240 (1970).
78. R. Hehlmann, B. Walter, N. Zollner, H. Wolf, and F. Deinhardt, "Infectious mononucleosis and acute monocytic leukemia," *Lancet* (Letter to the Editor) **2:**652–653 (1980).
79. J.E. Robinson, N. Brown, W. Andiman, K. Halliday, V. Francke, M.F. Robert, M. Andersson-Anvret, D. Horstmann, and G. Miller, "Diffuse polyclonal B-cell lymphoma during primary infection with Epstein-Barr virus," *N. Engl. J. Med.* **302:**1293–1297 (1980).
80. J.F. Fraumeni, Jr., "Infectious mononucleosis and acute leukemia," *J. Am. Med. Assoc.* (Letter to the Editor) **215:**1159 (1971).
81. G. Miller, T. Shope, L. Heston, R. O'Brien, A. Schwartz, and H. Pearson, "Prospective study of Epstein-Barr virus infections in acute lymphoblastic leukemia of children," *J. Pediatr.* **80:**932–937 (1972).
82. W.A. Andiman, "The Epstein-Barr virus and EB virus infections in childhood," *J. Pediatr.* **95:**171–182 (1979).

83. M.S. Reitz, B.J. Poiesz, F.W. Ruscetti, and R.C. Gallo, "Characterization and distribution of nucleic acid sequences of a novel type C retrovirus isolated from neoplastic human T lymphocytes," *Proc. Natl. Acad. Sci. USA* **78:**1887–1891 (1981).
84. The T- and B-Cell Malignancy Study Group, "Statistical analysis of immunologic, clinical and histopathologic data on lymphoid malignancies in Japan," *Jpn. J. Clin. Oncol.* **11:**15–38 (1981).
85. K. Takatsuki, T. Uchiyama, K. Sagawa, and J. Yodoi, in *Topics in Hematology,* pp. 73–77. (S. Seno, F. Takaku, and S. Irino, eds). Amsterdam: Excerpta Medica, 1977.
86. K. Takatsuki, T. Uchiyama, Y. Ueshima, and T. Hatori, "Adult T-cell leukemia: further clinical observations and cytogenetic and functional studies of leukemia cells," *Jpn. J. Clin. Oncol.* **9:**317–324 (1979).
87. T. Uchiyama, J. Yodoi, K. Sagawa, K. Takatsuki, and H. Uchino, "Adult T-cell leukemia: clinical and hematologic features of 16 cases," *Blood* **50:**481–492 (1977).
88. M. Robert-Guroff, Y. Nakao, K. Notake, Y. Ito, A. Sliski, and R.C. Gallo, "Natural antibodies to human retrovirus HTLV in a cluster of Japanese patients with adult T-cell leukemia," *Science* **215:**975–978 (1982).
89. V.S. Kalyanaraman, M.G. Sarngadharan, Y. Nakao, Y. Ito, T. Aoki, and R.C. Gallo, "Natural antibodies to the structural core protein (p24) of the human T-cell leukemia (lymphoma) retrovirus (HTLV) found in sera of leukemic patients in Japan," *Proc. Natl. Acad. Sci. USA* **79:**1653–1657 (1982).
90. V.S. Kalyanaraman, M.G. Sarngadharan, P.A. Bunn, J.D. Minna, and R.C. Gallo, "Antibodies in human sera reactive against an internal structural protein of human T-cell lymphoma virus," *Nature* **294:**271–273 (1981).
91. D. Catovsky, M.F. Greaves, M. Rox, D. Galton, A. Goolden, D. McCluskey, J. White, I. Lampert, G. Bourikas, R. Ireland, A. Brownell, J. Bridges, W. Blattner, and R. Gallo, "Adult T-cell lymphoma-leukemia in blacks from the West Indies," *Lancet* **1:**639–643 (1982).
92. I.I. Kessler and A.M. Lilienfeld, "Perspectives in the epidemiology of leukemia," *Adv. Cancer Res.* **12:**225–302 (1969).
93. G. Knox, "The epidemiology of childhood leukemia in Northumberland and Durham," *Br. J. Prev. Soc. Med.* **18:**17–24 (1964).
94. S.S. Meighan and G. Knox, "Leukemia in childhood. Epidemiology in Oregon," *Cancer* **18:**811–814 (1965).
95. D. Mainwaring, "Epidemiology of acute leukaemia of childhood in the Liverpool area," *Br. J. Prev. Soc. Med.* **20:**189–194 (1966).

96. F.W. Gunz and G.F. Spears, "Distribution of acute leukemia in time and space. Studies in New Zealand," *Br. Med. J.* **4:**604–608 (1968).
97. K. Pearson, "On the appearance of multiple cases of disease in the same house," *Biometrika* **8:**404–412 (1912).
98. K. Pearson, "Multiple cases of disease in the same house," *Biometrika* **9:**28–33 (1913).
99. F. Ederer, M.H. Myers, and N. Mantel, "A statistical problem in space and time: do leukemia cases come in clusters?" *Biometrics* **20:**626–638 (1964).
100. F. Ederer, M.H. Myers, H. Eisenberg, and P.C. Campbell, "Temporal-spatial distribution of leukemia and lymphoma in Connecticut," *J. Natl. Cancer Inst.* **35:**625–629 (1965).
101. D.E. Barton, F.N. David, and M. Merrington, "A criteria for testing contagion in time and space," *Ann. Hum. Genet. (London)* **29:**97–102 (1965).
102. F.N. David and D.E. Barton, "Two space-time interaction tests for epidemicity," *Br. J. Prev. Soc. Med.* **20:**44–48 (1966).
103. N. Mantel, "The detection of disease clustering and a generalized regression approach," *Cancer Res.* **27:**209–222 (1967).
104. M.R. Klauber, "Two-sample randomization tests of space-time clustering," *Biometrics* **27:**129–142 (1971).
105. M.C. Pike and P.G. Smith, "Disease clustering: a generalization of Knox's approach to the detection of space-time interactions," *Biometrics* **24:**541–556 (1968).
106. P. Greenwald, J.S. Rose, and P.B. Daitch, "Acquaintance networks among leukemia and lymphoma patients," *Am. J. Epidemiol.* **110:**162–177 (1979).
107. M.C. Pike and P.G. Smith, "A case-control approach to examine disease for evidence of contagion, including diseases with long latent periods," *Biometrics* **30:**263–279 (1974).
108. P.G. Smith and M.C. Pike, "Space-time clustering: a case-control method of examining diseases with long latent periods," in *Proceedings of the IARC/NCI Conference on "Host Environment Interactions in the Etiology of Cancer in Man."* Lyon, France: I.A.R.C., 1972.
109. N. Mantel and J.C. Bailar III, "A class of permutational and multinomial tests arising in epidemiologic research," *Biometrics* **26:**687–700 (1970).
110. N. Mantel and R. Valand, "A technique of nonparametric multivariate analysis," *Biometrics* **26:**547–558 (1970).
111. C.W. Heath and R.J. Hasterlik, "Leukemia among children in a suburban community," *Am. J. Med.* **34:**796–812 (1963).
112. C.W. Heath, Jr., M.D. Manning, and L. Zelkowitz, "Case clusters

in the occurrence of leukemia and congenital malformations," *Lancet* **2**:136–137 (1964).

113. H.R. Gilmore, Jr. and G. Zelesnick, "Environmental Hodgkin's disease and leukemia," *Pa. Med.* **65**:1047–1049 (1962).
114. P. McPhedran and C.W. Heath, Jr., "Multiple cases of leukemia associated with one house," *J. Am. Med. Assoc.* **209**:2021–2025 (1969).
115. I. Kemmoona, "Direct-contact clusters of acute lymphatic leukemia," *Lancet* (Letter to the Editor) **1**:994 (1974).
116. D.C. Bartsch, F. Springher, and H. Falk, "Acute nonlymphatic leukemia. An adult cluster," *J. Am. Med. Assoc.* **232**:1333–1336 (1975).
117. S.D. Reid, T.M. Hall, C.M. Reid, and P.J. Hull, "Cluster of myeloid leukaemia in Lytham St. Annes," *Lancet* (Letter to the Editor) **2**:579 (1979).
118. D. Pinkel and D. Nefzger, "Some epidemiological features of childhood leukemia in the Buffalo, N.Y. area," *Cancer* **12**:351–358 (1959).
119. D. Pinkel, J.E. Dowd, and I.D.J. Bross, "Some epidemiological features of malignant, solid tumors in the Buffalo, New York area," *Cancer* **16**:28–33 (1963).
120. P. Mustacchi, "Some intra-city variations of leukemia incidence in San Francisco," *Cancer* **18**:362–368 (1965).
121. F.E. Lundin, Jr., J.F. Fraumeni, Jr., J.W. Lloyd, and E.M. Smith, "Temporal relationships of leukemia and lymphoma deaths in neighborhoods," *J. Natl. Cancer Inst.* **37**:123–133 (1966).
122. J.F. Fraumeni, Jr., F. Ederer, and V.H. Handy, "Temporal-spatial distribution of childhood leukemia in New York State," *Cancer* **19**:996–1000 (1966).
123. M.M. Till, R.M. Hardisty, M.C. Pike, and R. Doll, "Childhood leukaemia in greater London: a search for evidence of clustering," *Br. Med. J.* **3**:755–758 (1967).
124. M.R. Klauber and P. Mustacchi, "Space-time clustering of childhood leukemia in San Francisco," *Cancer Res.* **30**:1969–1973 (1970).
125. M. Zahalkova, O. Bilek, A. Kubikova, and M. Belusa, "The incidence of leukemias in children, and the clustering in space and time in South Moravian Province, Czechoslovakia," *Blut* **21**:180–185 (1970).
126. R.J. Larsen, C.L. Holmes, and C.W. Heath, Jr., "A statistical test for measuring unimodal clustering: a description of the test and its application to cases of acute leukemia in metro Atlanta, Georgia," *Biometrics* **29**:301–309 (1973).

127. P.G. Smith, M. Pike, M.M. Till, and R.M. Hardistry, "Epidemiology of childhood leukaemia in Greater London: a search for evidence of transmission assuming a long latent period," *Br. J. Cancer* **33**:1–8 (1976).
128. C.R. Stark and N. Mantel, "Temporal-spatial distribution of birth dates for Michigan children with leukemia," *Cancer Res.* **27**:1749–1755 (1967).
129. M.R. Klauber, "A study of clustering of childhood leukemia by hospital of birth," *Cancer Res.* **28**:1790–1792 (1968).
130. A.G. Glass and N. Mantel, "Lack of space-time clustering of childhood leukemia in Los Angeles County, 1960–64," *Cancer Res.* **29**:1995–2001 (1969).
131. J.C. Bailar, H. Eisenberg, and N. Mantel, "Time between pairs of leukemia cases," *Cancer* **25**:1301–1303 (1970).
132. A.G. Glass, N. Mantel, F.W. Gunz, and G.F.S. Spears, "Time-space clustering of childhood leukemia in New Zealand," *J. Natl. Cancer Inst.* **47**:329–336 (1971).
133. A. Alperovitch, C. Hesse, P. Lazar, M. Hayat, and R. Flamant, "Temporal-spatial distributions of leukaemia and haematosarcoma in 7 French regions," *Int. J. Epidemiol.* **3**:209–218 (1974).
134. M.M. Zack, C.W. Heath, Jr., M.D. Andrews, A.S. Grivas, and B.W. Christine, "High school contact among persons with leukemia and lymphoma," *J. Natl. Cancer Inst.* **59**:1343–1349 (1977).
135. J. Guasch, "Hérédité des leucémies," *Le Sang* **25**:384–421 (1954).
136. D.M. Hayes, "The seasonal incidence of acute leukemia," *Cancer* **14**:1301–1305 (1961).
137. J.F. Fraumeni, Jr. "Seasonal variation in leukemia incidence," *Br. Med. J.* (Letter to the Editor) **2**:1408–1409 (1963).
138. A. Swan, "Seasonal variation in leukemia incidence," *Br. Med. J.* (Letter to the Editor) **2**:1063–1064 (1963).
139. J.A.H. Lee, "Seasonal variation in the clinical onset of leukemia in young people," *Br. Med. J.* **1**:1737–1738 (1962).
140. J.A.H. Lee, "Seasonal variation in leukaemia incidence," *Br. Med. J.* (Letter to the Editor) **2**:623 (1963).
141. J.A.H. Lee, "Seasonal variation in the incidence of the clinical onset of leukemia," *Pathol. Microbiol.* **27**:772–776 (1964).
142. D. Mainwaring, "Epidemiology of acute leukaemia of childhood in the Liverpool area," *Br. J. Prev. Soc. Med.* **20**:189–194 (1966).
143. E. von Ambs, J. Ströder and M. Ferneding, "Biostatische probleme akuter leukämien im kindesalter," *Arch. Kinderheilkd.* **173**:149–159 (1966).
144. A.G. Steinberg, "The genetics of acute leukemia in children," *Cancer* **13**:985–999 (1960).

145. A.M. Walker and P.A.H. van Noord, "No seasonality in the diagnosis of aute leukemia in the United States," *JNCI* **69:**1283–1287 (1982).
146. J.C. Bailar III and J.M. Gurian, "Month of birth and cancer mortality," *J. Natl. Cancer Inst.* **33:**237–242 (1964).
147. A.M. Stewart and D. Hewitt, "Aetiology of childhood leukemia," *Lancet* **2:**789–790 (1965).
148. F. Ederer, R.W. Miller, J. Scotto, and J.C. Bailar III, "Birth-month and infant cancer mortality," *Lancet* (Letter to the Editor) **2:**185–186 (1965).
149. J. Fedrick and E.D. Alberman, "Reported influenza in pregnancy and subsequent cancer in the child," *Br. Med. J.* **2:**485–488 (1972).
150. T. Hakulinen, L. Hovi, M. Karkinen-Jääskeläinen, A. Penttinen, and L. Saxen, "Associations between influenza during pregnancy and childhood leukaemia," *Br. Med. J.* **4:**265–267 (1973).
151. N.J. Vianna and A.K. Polan, "Childhood lymphatic leukemia: prenatal seasonality and possible association with congenital varicella," *Am. J. Epidemiol.* **103:**321–332 (1976).
152. A.M. Adelstein and J.W. Donovan, "Malignant disease in children whose mothers had chickenpox, mumps, or rubella in pregnancy," *Br. Med. J.* **4:**629–631 (1972).
153. V.L. Randolph and C.W. Heath, Jr., "Influenza during pregnancy in relation to subsequent childhood leukemia and lymphoma," *Am. J. Epidemiol.* **100:**399–409 (1974).
154. I. Leck and J.K. Steward, "Incidence of neoplasms in children born after influenza epidemics," *Br. Med. J.* **4:**631–634 (1972).

8 Immunologic Factors

Evidence indicating that the immune system must be, in one or more ways, related to the development of the leukemias is as follows:

1. Persons with immunodeficiency diseases, including X-linked (Bruton's) agammaglobulinemia, severe combined immunodeficiency, ataxia telangiectasia, and the Wiskott–Aldrich syndrome are at high risk for development of leukemia [1, 2].
2. Various drugs, including alkylating agents and other types of chemotherapy are associated with an excess incidence of leukemia, particularly ANLL, believed to result from the immunosuppressive activity of these agents (see Chapter 9).
3. Bacillus Calmette–Guerin (BCG) vaccination, widely used for human tuberculosis and shown to be an immunologic adjuvant and antitumor agent, may significantly reduce incidence or mortality from leukemia.
4. Chronic antigenic stimulation from a variety of autoimmune and/or inflammatory conditions is associated with an increased risk of CLL.

For several years, investigators have noted that children and adults with different primary immunodeficiency diseases are at increased risk for a number of malignancies. Since 1971, an Immunodeficiency Cancer Registry has collected data on individuals reported to the registry from the U.S. and abroad, and from the literature. In 1975, Spector summarized over 200 reported cases, which included 7 cases of leukemia among 12 with X-linked agammaglobulinemia reported to the registry, 5 out of 11 with severe combined immunodeficiency, 16 out of 70 with ataxia telangiectasia, and 2 of 34 with Wiskott–Aldrich syndrome. However, the cell types of leukemia were not reported, nor the risk of leukemia for populations with any of these rare conditions. Because the total number of individuals with such conditions is very small (incidence rates are not published), the occurrence of the leukemias, which are also rare, in persons with primary immunodeficiency diseases must be quite unlikely to have occurred by mere chance [1, 2].

Two studies published around 1970 showed a significant association between BCG vaccination and decreased mortality from leukemia. Davignon et al. noted that BCG vaccination of children in the province of Quebec subsequently decreased the death rate from leukemia in children under age 15 by a 2.4-fold factor. The leukemia death rate in the nonvaccinated group was reported to be 55.87 per 100,000 per year and in the vaccinated group, 23.54 per 100,000 per year [3]. Rosenthal et al. examined leukemia death rates in black children ages 0–6 years in Chicago, comparing rates in 172,986 unvaccinated black children with those in 54,414 vaccinated black children living in the same area of Chicago as the unvaccinated children, by searching death certificates from 1964–69. The vaccinated group received BCG at the age of 2–3 days while in the maternity division of Cook County Hospital. There were 21 leukemia deaths in the unvaccinated group. Rosenthal et al. calculated that this was consistent with a leukemia mortality rate of 2.02 per 100,000 per year. In contrast, the death rate for the vaccinated group, based on one death, was 0.31 per 100,000 per year. The difference was significant ($p = 0.04$) [4].

Both studies have been criticized. Stewart and Draper noted that vaccination status might have been misclassified in the Davignon et al. study since the 287 leukemia deaths must have been matched against one million names in the Central BCG Record. A child who was actually vaccinated may have been classified as unvaccinated. These authors believed that Davignon et al. did not consider the generally younger age distribution of children dying of leukemia compared to the older age of children receiving BCG vaccination in their analysis. Stewart and Draper pointed out that at least 70% of the vaccinations were carried out in children over age 5, whereas 40% of leukemia deaths in children under age 15 years occur before age 5. They suggested that the problem might be rectified by determining at what age the vaccinated children who developed leukemia had received BCG. Another discrepancy noted was that 20% of the children in Quebec had been vaccinated within a year of birth whereas 75% of the vaccinated leukemia cases had been vaccinated at birth [5, 6].

Hoover also described several methodologic problems in the Davignon et al. and the Rosenthal et al. studies. In both, death certificates were matched only with a roster of names of those vaccinated; this could allow for errors and potential misclassification. Both studies examined a vaccinated group that had received BCG as part of a service program. Controls had either not been included in the program or had refused to participate or had immigrated into the area. A randomized clinical trial would have been a better methodologic approach. Analysis did not involve an appropriate denominator; it failed to use the concept of person-years at risk. Hoover reanalyzed the Chicago data, with an appropriate denominator and found that the apparent protection conferred by BCG was a 50% rather than an 80%–85% reduction in leukemia mortality [7].

Kinlen and Pike compared the leukemia mortality in Quebec Province, where a large proportion of children ages 1–9 years had been vaccinated with BCG, with leukemia mortality in the rest of Canada, where a very small proportion of children had received BCG. For each of the age groups examined (0–4, 5–9, 10–14, and 15–19 years), leukemia mortality was no lower in Quebec than in the rest of Canada [8].

Negative evidence of a relationship has been reported. Leukemia rates (for all ages over 5 years) were examined by Comstock et al. in a vaccinated and unvaccinated group in Muscogee County, Georgia, and Russell County, Alabama, where vaccination had been part of a controlled trial. They found no differences in leukemia death rates in a 21-year follow-up of 16,913 vaccinated and 17,854 unvaccinated persons [9]. Snider et al. also found no significant differences in leukemia nor of other cancer incidence rates between 50,634 vaccinated and 27,338 nonvaccinated persons in a 23-year follow-up of a population in Puerto Rico who had participated in a control trial of BCG administration. However, analysis of Hodgkin's disease and lymphosarcoma combined incidence showed an excessive rate among the vaccinees [10]. Differences in results from those of Rosenthal et al. and Davignon et al. were attributed to possible variation in the antituberculosis (and perhaps in the antitumor) potency of the BCG vaccine which they used. Variation in vaccine efficacy has been shown for BCG compared to the vole bacillus vaccine [11]. Snider et al. noted that BCG administration had resulted in only a modest reduction of tuberculosis in this population (29% reduction). In general, the relationship between antituberculosis and antitumor potency is unknown. (Studies that showed a protective effect of BCG against leukemia, for the most part, consisted of vaccinees who had received BCG in the neonatal period.)

In an analysis of leukemia death rates, Waaler found differences in Sweden compared with Denmark and Norway. In Sweden, newborns were vaccinated with BCG whereas the vaccination was administered at age 7 in Denmark, at the time of school entrance, and at ages 13–14 in Norway. The lowest childhood leukemia death rates for Sweden were in the 0–4 age group whereas the lowest in Denmark were in the 5–9 age group and in Norway in the 15–19 year olds. However, Waaler points out that the differences are quite small, the numbers of deaths on which they are based are small and that these differences could be the result of random fluctuations. He concludes that additional study is needed [12].

A recurrent theme in the clinical literature is that lymphoproliferative malignancies are the result of chronic antigenic stimulation from a variety of sources [13–17]. This reflects a long-held general hypothesis about the etiology of CLL and plasma cell dyscrasias. The M-component abnormality sometimes seen in these conditions may reflect an excessive reticuloendothelial response to a specific stimulus, particularly one which is unusually protracted and resistant to elimination by the cellular and humoral reactions which it evokes. There are many anecdotal case reports of patients developing a lymphoproliferative malignancy such as CLL or plasma cell dyscrasias following a long history of osteomyelitis, tuberculosis, pyelonephritis, biliary tract disease, chronic bronchitis, diverticulitis, syphilis, chronic sinusitis, chronic pancreatitis, and chronic dermatitis [13, 14, 16, 17]. However, there is no reported epidemiological study which examined the associations between chronic infections or inflammatory processes and CLL.

In a study of the incidence and intensity of schistosomiasis in Ibadan in the Western State of Nigeria, Edington et al. [18] examined the bladders of 673 autopsied persons as well as the rectum, kidney, liver, lung, and spleen. The pathologic findings in cadavers with schistosomiasis and in noninfected cadavers were compared. A significantly higher proportion of lymphoproliferative malignancies, except for Burkitt's tumor, was noted in cases with schistosome infection. The frequency of the lymphoproliferative tumors was unrelated to the number of eggs in the bladder. Similarly, in Egypt, a high frequency of lymphoproliferative tumors has been associated with schistosomiasis [15]. Further confirmation is supplied by the reported occurrence of reticulum cell sarcoma in experimental schistosomiasis. Of course, there are alternative explanations for the association other than antigenic stimulation. One possibility is that drugs used for treatment for schistosomiasis may be responsible. In listing the types of lymphoproliferative malignancies, the authors combined all leukemias, so it is not possible to determine if there was an increased incidence of CLL.

Several studies noted associations between several autoimmune diseases and subsequent development of lymphocytic or plasma cell neoplasms. Most of these reports have causally linked chronic antigenic stimulation from diseases such as rheumatoid arthritis [19, 20], systemic lupus erythematosus, dermatomyositis [21], Sjögren's syndrome [22], or a variety of other diseases [21, 23] with subsequent development of chronic lymphocytic leukemia, malignant lymphoma, or other lymphoproliferative disorders. Blackburn et al. carried out a follow-up study of 1,625 persons with pernicious anemia in the United Kingdom. Between 1959 and 1963, there was a 3.3-fold excess number of deaths from leukemia in this population [24]. Abbott and Lea compared the past medical histories in records of 679 leukemics with those of controls having other diseases. An excess of the leukemics had a history of rheumatic diseases compared with controls [25]. These same investigators compared the medical records of 734 persons with leukemia, identified from the Armed Forces, with age-, sex-, and calendar-period-matched controls. For the leukemics, 10.9% had given a history of rheumatic disease compared with 3.7% of the controls [26].

Vianna and Bross found associations between several chronic (at least a five year history) autoimmune and inflammatory conditions in their analysis of the Tri-State leukemia case control study, comparing 1,345 leukemics and 1,237 random controls. Elevated risks were found for the association between psoriasis and lymphatic leukemia in men (R.R. = 5.06), tuberculosis and lymphatic leukemia in men (R.R. = 5.54) and myeloid leukemia in women (R.R. = 8.63), "rheumatism" and lymphatic leukemia in men (R.R. = 3.12), "rheumatism" and myeloid leukemia in women (R.R. = 2.94), and eczema and lymphatic leukemia (R.R. = 4.82) and myeloid (R.R. = 7.07) leukemia in men [27].

Various investigators have considered the role of chronic antigenic stimulation in the development of lymphoproliferative malignancies [13]. Thus far, they have been unable to identify specific antigens that can be shown to reliably reproduce CLL or other lymphoproliferative malignancies in animals or man, with the exception of mineral-oil-induced plasma cell tumors in BALB/c mice.

Potter [28] has shown that the cells of transplantable or primary plasma cell tumors (a type of lymphoproliferative malignancy) in BALB/c mice synthesize and secrete only one type of immunoglobulin molecule. These homogeneous immunoglobulins have been shown to bind specific antigens including purine and pyrimidine-substituted proteins, pneumococcal polysaccharides, streptolysin, serum lipoprotein, red blood cell antigens, and gammaglobulins. He has shown that the precursors of neoplastic plasma cells are actively forming antibodies to immunogens that originate in microorganisms that inhabit the gastrointestinal tract of BALB/c mice. Specifically, he found that the immunogens were heteropolysaccharides derived from bacteria of the normal enteric flora. These animal data are suggestive but it is not possible to extrapolate these findings to CLL or plasma cell tumors in man without additional evidence. The epidemiologic approach may provide a method by which one could investigate some of the implications of Potter's work for related tumors in man.

Four lines of evidence are suggestive of a causal association between immunologic factors and the leukemias. First, a significant proportion of persons with a variety of rare primary immunodeficiency disorders have been reported to subsequently develop leukemia, although quantitative measures of risk and cell types of leukemia have not been determined. Another large body of work, discussed in detail in Chapter 9, has demonstrated that a variety of immunosuppressive drugs may be leukemogenic, particularly for ANLL. A third, but somewhat weaker line of evidence suggests that BCG vaccination, with immunologic adjuvant and antitumor properties, may be protective against leukemia development. Finally, chronic antigenic stimulation as provided by a number of different chronic infectious, inflammatory, and autoimmune disorders may be associated with an excess risk of CLL, although most of the current evidence supporting this association consists of case reports and series, rather than case-control comparisons.

REFERENCES

1. J.H. Kersey, B.D. Spector, and R.A. Good, "Primary immunodeficiency diseases and cancer: the Immunodeficiency Cancer Registry," *Int. J. Cancer* **12:**333–347 (1973).
2. B.D. Spector, "Immunodeficiency Cancer Registry: 1975 update," in *Genetics of Human Cancer,* pp. 339–342. (J.J. Mulvihill, R.W. Miller, and J.F. Fraumeni, Jr., eds). New York: Raven Press, 1977.
3. L. Davignon, P. Lemonde, P. Robillard, and A. Frappier, "BCG vaccination and leukaemia mortality," *Lancet* **2:**638 (1970).
4. S.R. Rosenthal, R.G. Crispen, M.G. Thorne, N. Piekarski, N. Raisys, and P.G. Rettig, "BCG vaccination and leukemia mortality," *J. Am. Med. Assoc.* **222:**1543–1544 (1972).
5. A.M. Stewart and G. Draper, "BCG vaccination and leukaemia mortality," *Lancet* (Letter to the Editor) **2:**983 (1970).
6. A.M. Stewart and G. Draper, "BCG vaccination and leukemia mortality," *Lancet* (Letter to the Editor) **1:**799 (1971).
7. R.N. Hoover, "Bacillus Calmette–Guérin vaccination and cancer prevention: a critical review of the human experience," *Cancer Res.* **36:**652–654 (1976).
8. L.J. Kinlen and M.C. Pike, "BCG vaccination and leukaemia. Evidence of vital statistics," *Lancet* **2:**398–402 (1971).
9. G.W. Comstock, I. Martinez, and V.T. Livesay, "Efficacy of BCG vaccination in prevention of cancer," *J. Natl. Cancer Inst.* **54:**835–839 (1975).
10. D.E. Snider, G.W. Comstock, I. Martinez, and G.J. Caras, "Efficacy of BCG vaccination in prevention of cancer: an update," *J. Natl. Cancer Inst.* **60:**785–788 (1978).
11. Medical Research Council, "BCG and vole bacillus vaccines in the

prevention of tuberculosis in adolescence and early adult life," *Bull. WHO* **46:**371–385 (1972).

12. H.T. Waaler, "BCG and leukemia mortality," *Lancet* (Letter to the Editor) **2:**1314 (1970).
13. W.A. Blattner, "Epidemiology of multiple myeloma and related plasma cell disorders: an analytic review," in *Progress in Myeloma,* pp. 1–65 (M. Potter, ed). New York: Elsevier/North Holland, 1980.
14. Combined Staff Clinic, "Plasma cell dyscrasias—current clinical and biochemical concepts," *Am. J. Med.* **44:**256–269 (1968).
15. M. El Gazayerli, M. Kharadly, R. Khalil, R. Galal, W. Riad, and M.M. Gazayerli, "Primary tumours of lymph nodes," in *The Lymphoreticular Tumours in Africa,* pp. 36–41 Symposium, International Union Against Cancer, Paris (F.C. Roulet, ed). Basel: S. Karger, 1963.
16. R.W. Rundles, "Chronic lymphocytic leukemia," in *Hematology,* 2nd ed, pp. 1002–1020 (W.J. Williams, E. Beutler, A.J. Ersley, and R.W. Rundles, eds., New York: McGraw-Hill, 1977.
17. Z.A. Zawadzki and G.A. Edwards, "Dysimmunoglobulinemia associated with hepatobiliary disorders," *Am. J. Med.* **48:**196–202 (1970).
18. G.M. Edington and C.M.V. Maclean, "The relative incidence of tumors of the reticulo-endothelial system in Ibadan, Nigeria," in *The Lymphoreticular Tumors in Africa,* pp. 54–66 Symposium, International Union Against Cancer, Paris (F.C. Roulet, ed). Basel: S. Karger, 1963.
19. C.J. Goldenberg, F. Daraskevas, and L.G. Israels, "The association of rheumatoid arthritis with plasma cell and lymphocyte neoplasms," *Arthritis Rheum.* **12:**569–576 (1969).
20. H.A. Isomaki, T. Hakulinen, and U. Joutsenlahti, "Excess risk of lymphoma, leukemia, and myeloma in patients with rheumatoid arthritis," *J. Chronic Dis.* **31:**691–696 (1978).
21. C.L. Conley, J. Misiti, and A.J. Laster, "Genetic factors predisposing to chronic lymphocytic leukemia and to autoimmune disease," *Medicine* **59:**323–334 (1980).
22. N. Talal and J.J. Bunim, "The development of malignant lymphoma in the course of Sjögren's syndrome," *Am. J. Med.* **36:**529–540 (1964).
23. A. Oleinick, "Leukemia or lymphoma occurring subsequent to an autoimmune disease," *Blood* **29:**144–153 (1967).
24. E.K. Blackburn, S.T. Callendar, J.V. Dacie, R. Doll, R.H. Girdwood, D.L. Mollin, R. Saracci, J.L. Stafford, R.B Thompson, S. Varadi, and G. Wetherly-Mein, "Possible association between pernicious anaemia and leukaemia: a prospective study of 1,625 patients with a note on the very high incidence of stomach cancer," *Int. J. Cancer* **3:**163–170 (1968).

25. J.D. Abbatt and A.J. Lea, "Leukaemogens," *Lancet* **2**:880–883 (1958).
26. A.J. Lea, "An association between the rheumatic diseases and the reticuloses," *Ann. Rheum. Dis.* **23**:480–484 (1964).
27. E. Vianna and I.D.J. Bross, "Leukemia prediction by medical history," *Prev. Med.* **3**:165–170 (1974).
28. M. Potter, "Mouse IgA myeloma proteins that bind polysaccaride antigens of enterobacterial origin," *Fed. Proc.* **29**:85–91 (1970).

9 Leukemia as a Secondary Malignancy in Persons with Malignant and Nonmalignant Diseases

Cases of leukemia as a second malignancy occurring before, simultaneously with, or following various other neoplasms have been reported for over one hundred years. In 1978, Whipham reported a case of leukemia coexistent with another type of malignant lesion [1, 2]. From 1911 through 1969, 16 cases of Hodgkin's disease were reported to have terminated in acute nonlymphocytic leukemia (ANLL) [3]. Since 1970, well over 150 such cases have been reported [3, 4]. During the past 10 to 15 years, ANLL as a second malignancy has been noted to occur with increased frequency in patients with multiple myeloma, non-Hodgkin's lymphoma, ovarian cancer, breast cancer, and lung cancer. Non-Hodgkin's lymphoma also has been reported to occur in excess in persons receiving prolonged immunosuppressive treatment, including kidney transplant recipients.

METHODOLOGIC CONSIDERATIONS

The increasing numbers of reports of cases of ANLL as a secondary malignancy in persons with malignant (and nonmalignant) diseases may not necessarily imply an increasing incidence. In order to determine the incidence of leukemia as a secondary malignancy, it is necessary to study and completely follow-up a well-defined population of cases with a malignancy, such as Hodgkin's disease or multiple myeloma. Another approach is to compare the cumulative probabilities of developing ANLL as a second neoplasm among successive historical cohorts of patients using life-table methods of analysis [5, 6]. An appropriate control group for which leukemia-specific incidence rates are available by age, race, and sex for the same calendar period is essential in order to determine the expected number of cases of ANLL. Rigorous criteria for definition and histopathologic classification of the second malignancy, ANLL, need to be employed to be certain that it is not a recurrence or metastasis of the primary malignancy. Also, an appropriate latent period between the onset of the first primary malignancy and the onset of ANLL must be defined to insure that ANLL is a metachronous or subsequent malignant neoplasm rather than a synchronous neoplasm or a previous neoplasm [5].

Adequate sample sizes and sufficient years of follow-up for both case and control groups are required to be able to detect significant differences. Case groups selected from different institutions must be similarly classified according to histopathology and stage of disease. Cases and controls must be carefully and comparably classified as to sociodemographic characteristics. The case population under study must also be well characterized in terms of referral patterns, sociodemographic factors, treatment parameters, treatment stage (i.e., initial treatment versus treatment for relapse). Finally, it is essential to take into account the fact that the risk of leukemia as a second malignancy may not remain constant over time because treatment modalities may change during the course of the disease.

ALKYLATING AGENTS

Evidence from animal studies indicates that several alkylating agents including procarbazine, nitrogen mustard, chlorambucil, melphalan, cyclophosphamide, actinomycin, adriamycin, and azathioprine display strong measurable carcinogenic activity [7–14]. Most of these alkylating agents as well as other antitumor agents also can induce tumors in laboratory animals. Procarbazine is one of the most potent carcinogens; it can induce a variety of tumors in rats, mice, and primates. However, carcinogenicity as shown in laboratory animals does not necessarily translate to carcinogenicity in man. Other explanations may account for the apparent increased incidence of leukemia in humans treated with alkylating and other antitumor agents. Various types of these chemotherapeutic drugs have been shown to suppress humoral and cellular immune function, to cause chromosomal abnormalities or to display mutagenicity.

Some cancer patients, including those with Hodgkin's disease, are at a higher risk of second malignancies due to host factors, such as immunogenetic characteristics, interaction between host and treatment factors, sensitivity of various tissues such as bone marrow, etc. Also, ANLL may be part of the natural history of various hematopoietic malignancies including Hodgkin's disease, multiple myeloma, and non-Hodgkin's lymphoma. The apparent increased incidence of ANLL as a second primary malignancy may be due to longer survival times which may have been a result of improvements in therapy.

However, epidemiologic evidence has indicated that the use of antineoplastic therapy, in addition to the other risk factors described above, may represent a separate independent risk factor for subsequent development of a secondary malignancy, particularly leukemia. Patients with various types of malignancies including Hodgkin's disease, multiple myeloma, breast cancer, ovarian cancer, and lung cancer have been shown to have increased risk of developing acute leukemia following therapy with antineoplastic drugs [15–21].

Some of these patients have been treated only with chemotherapy or with radiation; others have received combined modality treatment of chemotherapy and radiotherapy. Ionizing radiation (see Chapter 6) has been shown to induce several types of neoplasms including ANLL. The dose-response curve appears to vary with the type of neoplasm, age at exposure to radiation, host factors (immunogenetic, hormonal, and other physiologic characteristics), dose of radiation, type of radiation, anatomic site of radiation administration, and other treatments used.

Several characteristics of the acute nonlymphocytic leukemias which occur as second malignancies suggest that they are a distinct clinicopathologic entity from ANLL as a primary malignancy [22]. The ANLL as a second malignancy is sometimes preceded by a sideroblastic anemia lasting for a variable period of time. Other times, the leukemia is preceded by the development of pancytopenia. As the leukemia develops, the marrow often has a megaloblastoid appearance; nucleated red blood cells are seen in the peripheral blood. Random chromosomal aberrations are seen in a majority of these leukemia bone marrows. There is usually a refractoriness to antileukemia therapy and survival is very short (median one month). While some of these characteristics are not unique, the constellation does somewhat differentiate this type of ANLL from acute nonlymphocytic leukemias which are first primaries.

Numerous investigators have reported the occurrence of ANLL as a second malignancy at different intervals following diagnosis of various cancers, particularly those treated with chemotherapy and/or radiotherapy [4, 15, 20–37]. However, the significance of case reports is difficult to interpret and excess risk is impossible to estimate in the absence of comparison groups. Therefore, our summary tables focus only on analytic studies.

A fairly representative summary of the analytic studies of leukemia as a second malignancy following Hodgkin's disease and multiple myeloma is shown in Tables 9-1 and 9-2. Nineteen studies of leukemia as a second malignancy in Hodgkin's disease patients are summarized in Table 9-1 [1, 6, 38–54]. The study group sizes ranged from 82 to 10,000 patients. Patients followed-up in the 19 studies were treated in a variety of ways

Table 9-1 Leukemia as a secondary malignancy in patients with Hodgkin's disease

Author; year pub; (yrs of study)	No. of cases	Leukemia type	No. of leukemia cases (% of cases observed)	Crude incidence of leukemia per 100,000 per year	Relative risk	Control population	Remission status at onset of leukemia
Moertel and Hagedorn, 1957; (1944–53) [1]	826	All Types	0	—	—	None	—
Peters and Middlemiss, 1958; (1928–54) [38]	291	AML	2 (0.68%)	154	—	None	Both in remission
Berg, JW, 1967; (1949–62) [39]	1028	AML	0	—	—	New York State population	—
Ezdinli et al., 1969; (1925–69) [40]	Not given. (3000 person-years of Hodgkin's disease observed)	Case #1 AML	3	100	15	US population (death rates, 1965)	#1 Partial remission
		Case #2 Erythroleukemia					#2 Complete remission
		Case #3 CML					#3 Complete remission
Crosby, 1969; (1950–69) [41]	10,000	AML	17 (0.17%)	—	10 (estimated)	US population leukemia incidence rates	—
Newman et al., 1970; (1950–67) [42]	1500	AMMoL (all 3 cases)	3 (0.20%)	—	"Exceptionally high"	None	#1 In complete remission #2 Hodgkin's disease present #3 Hodgkin's disease present
Weiden et al., 1973; (1961–72) [43]	160	ANLL	3 (1.9%)	—	—	None	#1 In complete remission

Chemotherapy assoc. with leukemia	Duration of treatment	Role of radiation[b]	Time between diagnosis of malignancy and diagnosis of leukemia	Age[a]/sex distribution of leukemia	Response of leukemia to treatment	Other
—	—	—	—	—	—	—
—	—	Both treated with this modality	—	—	—	—
—	—	—	—	—	—	—
#1 First received chemotherapy 1 month before leukemia diagnosed	—	#1 received extensive radiation intermittently for 7 years	#1 7 years	#1 Female Age 33	#1 None	—
#2 Never treated with chemotherapy	—	#2 received radiation over a 6-month period	#2 2 years	#2 Male Age 72	#2 None	—
#3 Vincristine received for 7 weeks in 1962	—	#3 received radiation 1963, 1966	#3 8 years	#3 Male Age 12	#3 Good 9/67–1/69	—
—	—	—	—	—	—	—
#1 Cortisone given initially intermittently; nitrogen mustard given 1 yr prior to diagnosis of leukemia	#1 16 years	#1 Administered 10 times	#1 18 years	#1 White male Age 9	#1 No response	—
#2 None	#2 1½ years	#2 Administered 4 times	#2 1½ years	#2 White male Age 25	#2 No response	—
#3 Nitrogen mustard and chlorambucil over a 5-year period	#3 5½ years	#3 Administered 8 times	#3 5½ years	#3 White female Age 38	#3 Treatment not given	—
#1 None	—	#1 Radiotherapy only	#1 3½ years	#1 White male Age 51	#1 Apparent response but died of complications of marrow transplant	—

Table 9-1 (continued)

Author; year pub; (yrs of study)	No. of cases	Leukemia type	No. of leukemia cases (% of cases observed)	Crude incidence of leukemia per 100,000 per year	Relative risk	Control population	Remission status at onset of leukemia
Weiden (continued)	160	ANLL	3 (1.9%)	—	—	None	#2 Residual Hodgkin's disease #3 —
Bonadonna et al., 1973; (1970–73) [44]	82	AML	2 (2.4%)	—	—	None	—
Sahakian, 1974; (1958–73) [45]	1095	#1 AMMoL	3 (0.3%)	—	—	None	#1 Residual Hodgkin's disease
		#2 AMMoL	—	—	—	—	#2 Residual Hodgkin's disease
		#3 AMoL	—	—	—	—	#3 Residual Hodgkin's disease
Newell et al., 1974; (1948–72) [46]	350	—	1 (0.29%)	—	—	US population rates from the Ten Cities Survey, 1947	—
Canellos, 1975; (not stated) [47]	452	#1 AML	2 (0.44%)	—	Not stated separately for leukemia	US population, Third National Cancer Survey, 1969–71	—
		#2 AMMoL	—				—
Jacquillat et al., 1976 (not stated) [48]	1770	AML	11 (0.62%)	—	—	None	—

Table 9-1 (continued)

Chemotherapy assoc. with leukemia	Duration of treatment	Role of radiation[b]	Time between diagnosis of malignancy and diagnosis of leukemia	Age[a]/sex distribution of leukemia	Response of leukemia to treatment	Other
#2 Treated with chlorambucil, vincristine, alkylating agents, procarbazine, and steroids	—	#2 Administered	#2 8 years	#2 White male Age 16	#2 —	—
#3 No chemotherapy given	—	#3 Total nodal irradiation administered	#3 1½ years	#3 White male Age 26	#3 Initially stabilized without therapy but expired 5 months later	—
—	—	—	—	—	—	—
#1 Initially treated with chlorambucil and vinblastine. At relapse treated with procarbazine and bleomycin	#1 Treated on and off over 9 years	#1 Administered	#1 8½ years	#1 White female Age 32	#1 No response	—
#2 MOPP and BCNU	#2 8 months	#2 None	#2 9 months	#2 White male Age 49	#2 No response	—
#3 Chlorambucil, procarbazine, vincristine, bleomycin	—	#3 Administered	#3 8 years	#3 White female Age 21	#3 No response	—
—	—	—	—	—	—	—
#1 MOPP following irradiation	#1 6 cycles	#1 Intensive radiation administered	#1 —	#1 Male Age 27	#1 —	—
#2 MOPP following irradiation	#2 6 cycles	#2 Intensive radiation administered	#2 —	#2 Female Age 19	#2 —	—
7/11 MOPP	6 cycles	Standard radiotherapy	5–11 years following diagnosis of Hodgkin's for all cases	—	—	—
2/11 Maintenance chemotherapy	—	Standard radiotherapy				
2/11 MOPP	6 cycles	Extended radiation				

Table 9-1 (continued)

Author; year pub; (yrs of study)	No. of cases	Leukemia type	No. of leukemia cases (% of cases observed)	Crude incidence of leukemia per 100,000 per year	Relative risk	Control population	Remission status at onset of leukemia
Larsen and Brinker, 1977; (1964–75) [49]	201	#1 AMMoL	3 (1.5%)	4.52	75	Danish population leukemia rates 1963–67	#1 Hodgkin's disease active
		#2 Erythroleukemia	—	—	—		#2 Hodgkin's disease in remission
		#3 AMMoL	—	—	—		#3 Hodgkin's disease in remission
Newell and Krementz, 1977; (1948–70) [50]	350	AML	1 (0.29%)	1.04	—	US Population Ten Cities Survey, 1947	—
Brody, 1977; (1950–54) [6]	366	—	0	—	—	Connecticut Cancer Registry	—
(1960–64)	314	#1 AMoL	2 (0.64%)	—	—		#1 —
		#2 AMoL	—	—	—		#2 —
(1968–72)	348	#1 AMMoL	1 (0.29%)	—	—		#1 —
Coleman, 1977; (1968–75) [51]	680	6/9 AML 1/9 acute undiff. leukemia 1/9 subacute leukemia 1/9 myeloproliferative syndrome	9 (1.3%)	—	—	None	6/9 in complete remission 3/9 had active Hodgkin's disease

Table 9-1 (continued)

Chemotherapy assoc. with leukemia	Duration of treatment	Role of radiation[b]	Time between diagnosis of malignancy and diagnosis of leukemia	Age[a]/sex distribution of leukemia	Response of leukemia to treatment	Other
#1 No chemotherapy	—	#1 Total nodal irradiation administered	#1 2 years	#1 White male Age 55	#1 No response	—
#2 Maintenance chemotherapy with cyclophosphamide, 3½ years later given MOPP. 2 years later MOPP given again	—	#2 Administered	#2 9 years	#2 White male Age 43	#2 No response	—
#3 MOPP given 2 years after diagnosis for relapse	—	#3 Administered	#3 6½ years	#3 White male Age 23	#3 Stable on conservative therapy	—
—	—	—	—	—	—	—
—	—	—	—	—	—	—
#1 —	#1 —	#1 —	#1 8⅓ years	#1 Female Age 23	#1 —	—
#2 —	#2 —	#2 —	#2 8 years	#2 Female Age 22	#2 —	—
#1 MOPP and BCNU followed by vinblastine & prednisone	6 cycles	#1 No radiation	#1 ⅝ years	# Male Age 48	#1 —	—
#1 MOPP	#1 6 cycles	#1 Initially treated with total nodal	#1 3 years	#1 Male Age 26	#1 None	—
#2 MOPP	#2 5 cycles	#2 Initially treated with total nodal	#2 4 years	#2 Male Age 29	#2 None	—
#3 MOPP	#3 3 cycles	#3 Initially treated with total nodal	#3 1½ years	#3 Female Age 19	#3 None	—
#4 MOPP	#4 6 cycles	#4 Initially treated with total nodal	#4 7½ years	#4 Female Age 27	#4 None	—
#5 MOPP	#5 6 cycles	#5 Initially treated with total nodal	#5 2¾ years	#5 Male Age 32	#5 None	—
#6 No chemotherapy following radiation for 2 years then additional radiation then MOPP	#6 6 cycles	#6 Initially treated with mantle irradiation	#6 2¾ years	#6 Male Age 41	#6 None	—

Table 9-1 (continued)

Author; year pub; (yrs of study)	No. of cases	Leukemia type	No. of leukemia cases (% of cases observed)	Crude incidence of leukemia per 100,000 per year	Relative risk	Control population	Remission status at onset of leukemia
Coleman (continued)	680		9 (1.3%)	—	—	None	
Toland et al., 1978; (not stated) [52]	643	—	11 (1.7%)	8.31	—	US Third National Cancer Survey incidence rates	10/11 in complete remission
Valagussa, 1980; (1965–76) [53]	764	5/9 AMMoL 4/9 AML	9 (1.2%)	—	—	None	—
Baccarani et al., 1980; (1969–76) [54]	613	4/7 AML 1/7 AMMoL 2/7 Erythroleukemia	7 (1.1%)	—	—	None	#1 In complete remission #2 In complete remission

Chemotherapy assoc. with leukemia	Duration of treatment	Role of radiation[b]	Time between diagnosis of malignancy and diagnosis of leukemia	Age[a]/sex distribution of leukemia	Response of leukemia to treatment	Other
#7 Intravenous colloidal gold	—	#7 Initially treated with total nodal	#7 4¼ years	#7 Male Age 56	#7 None	—
#8 MOPP followed by single agent chemotherapy then brain irradiation	#8 6 cycles	#8 Initially treated with mantle	#8 5 years	#8 Male Age 39	#8 None	—
#9 No chemotherapy	#9 Total nodal irradiation	#9 Initially treated with total nodal	#9 3¾ years	#9 Female Age 12	#9 None	—
3/11 Treated with no radiotherapy; given intensive chemotherapy	At least 6 mos. chemo.	—	—	—	—	*"DOSE"* *R.R.* No radiotherapy (RT) plus intensive chemotherapy (IC) 257
2/11 Treated with some radiotherapy and intensive chemotherapy	< TNI or extended field RT plus > 6 mos. chemo.	—	—	—	—	Some RT plus IC 253
4/11 Treated with intensive radiotherapy and some chemotherapy	TNI or extended field RT plus < 6 mos. chemo.	—	—	—	—	Intensive RT and some chemotherapy 816
2/11 Treated with intensive radiotherapy and intensive chemotherapy	TNI or extended field RT plus > 6 mos. chemo.	—	—	—	—	Intensive RT and IC 1049 (A dose-response effect is seen)
5/9 treated with MOPP following radiotherapy	—	8/9 treated with radiation	5/9 / 2½–12½ yrs	—	1/9 Responded to treatment 8/9 Showed no response to treatment	—
2/9 Treated with MABOP (adriamycin and bleomycin substituted for procarbazine) following radiotherapy			2/9 / 2¾ yrs 5⅓ yrs			
1/9 Treated with mustard and procarbazine following radiotherapy			1/9 / 8½ yrs			
1/9 Treated with MABOP only			1/9 / 1⅔ yrs			
#1 MOPP (3 cycles) and Procarbazine	—	#1 Total nodal irradiation (TNI) administered	#1 3½ years	#1 Male Age 29	#1 No response	Actuarial frequency of leukemia 7 years after diagnosis of Hodgkin's disease was 2.0% for those treated with chemotherapy and 2.04%
#2 Chlorambucil and prednisone	—	#2 TNI administered	#2 7 years	#2 Female Age 30	#2 No response	

Table 9-1 (continued)

Author; year pub; (yrs of study)	No. of cases	Leukemia type	No. of leukemia cases (% of cases observed)	Crude incidence of leukemia per 100,000 per year	Relative risk	Control population	Remission status at onset of leukemia
Baccarani (continued)	613		7 (1.1%)	—	—	None	#3 In complete remission #4 In complete remission #5 In complete remission #6 In complete remission #7 Active Hodgkin's disease

[a]At the time of diagnosis of primary malignancy.
[b]Radiation dose, frequency, port often not described.

Table 9-2 Leukemia as a secondary malignancy in patients with multiple myeloma

Author; year pub; (yrs of study)	No. of cases	Leukemia type	No. of leukemia cases (% of cases observed)	Crude rate of ANLL per 1,000 person-years	Relative risk	Control population	Remission status at onset of leukemia
Weitzel, 1958; (1930–56) [55]	57 (autopsy series)	ANLL	0	—	—	—	—
Nordenson, 1966; (1932–63) [56]	310	ANLL	2 (0.6%)	—	—	None	—
Edwards and Zawadzki, 1967; (1950–66) [57]	78	AML	1 (1.3%)	—	—	None	Multiple myeloma active
Kyle et al., 1970; (1964–66) [58]	250	AMMoL	3 (1.2%)	—	—	—	#1 Partial remission #2 Partial remission #3 Complete remission

Chemotherapy assoc. with leukemia	Duration of treatment	Role of radiation[b]	Time between diagnosis of malignancy and diagnosis of leukemia	Age[a]/sex distribution of leukemia	Response of leukemia to treatment	Other
#3 MOPP (6) and cytoxan and procarbazine		#3 TNI administered	#3 4½ years	#3 Male Age 43	#3 No response	for those treated with chemotherapy and radiotherapy
#4 MOPP (14) and BCG	—	#4 TNI administered	#4 5 years	#4 Male Age 47	#4 No response	
#5 MOPP (6)	—	#5 TNI administered	#5 2½ years	#5 Female Age 26	#5 No response	
#6 MOPP (4)	— —	#6 TNI administered	#6 3 years	#6 Male Age 40	#6 Supportive care only—12 months survival	
#7 Vinblastine and bleomycin and MOPP (1)		#7 TNI administered		#7 Male Age 63	#7 No response	

Chemotherapy assoc. with leukemia	Duration of treatment	Role of radiation	Time between diagnosis of malignancy and diagnosis of leukemia	Age[a]/sex distribution of leukemia	Response of leukemia to treatment	Actuarial risk
—	—	—	—	—	—	—
—	—	—	—	—	—	—
Initially urethane, 8 years later cyclophosphamide	—	Radiation administered 3 years prior to development of leukemia	10 years	Age 62 White male	No response	—
#1 Melphalan	#1 3¾ years	#1 Administered to spine and R shoulder	#1 4 years	#1 Male Age 55	#1 No response	—
#2 Melphalan	#2 2⅝ years	#2 None	#2 2⅝ years	#2 Male Age 54	#2 No response	
#3 Melphalan	#3 2½ years	#3 1200 rads given to spine	#3 2½ years	#3 Male Age 67	#3 No response	

Table 9-2 (continued)

Author; year pub; (yrs of study)	No. of cases	Leukemia type	No. of leukemia cases (% of cases observed)	Crude rate of ANLL per 1,000 person-years	Relative risk	Control population	Remission status at onset of leukemia
Anderson and Videbaek, 1970; (1960–64) [59]	19	AML	4 (2.1%)	—	—	—	#1 Multiple myeloma active #2 Multiple myeloma active #3 Multiple myeloma active #4 Partial remission
Khaleeli et al., 1973; (1962–72) [60]	75	AMMoL	4 (5.3%)	—	—	—	#1 Complete remission #2 — #3 — #4 —
Karchner et al., 1974; (1966–72) [61]	63	AMoL AML AML AML AMMoL	5 (7.9%)	—	29	New York State Cancer Registry (to calculate NYS incidence rates)	#1 — #2 — #3 — #4 — #5 —
Newell et al., 1974; (1948–70) [46]	261	ANLL	0	—	—	—	—

Chemotherapy assoc. with leukemia	Duration of treatment	Role of radiation	Time between diagnosis of malignancy and diagnosis of leukemia	Age[a]/sex distribution of leukemia	Response of leukemia to treatment	Actuarial risk
#1 Melphalan	#1 2½ years	#1 None	#1 3½ years	#1 Female Age 82	#1 No response	—
#2 Melphalan	#2 1½ years	#2 RT administered	#2 1½ years	#2 Female Age 71	#2 No response	
#3 Melphalan	#3 4 years then cyclophosphamide	#3 None	#3 7 years	#3 Male Age 65	#3 No response	
#4 Prednisone and cyclophosphamide	#4 5 years	#4 RT administered	#4 5½ years	#4 Female Age 61	#4 No response	
#1 Melphalan	#1 7 years	#1 None	#1 8 years	#1 White male Age 48	#1 No response	—
#2 Melphalan	#2 8½ years	#2 None	#2 9 years	#2 White male Age 50	#2 No remission but survived 8 months following diagnosis	
#3 Melphalan	#3 3¼ years	#3 None	#3 4 years	#3 White female Age 55	#3 Partial remission on cyclophosphamide, vincristine, cytosine arabinoside, and prednisone followed by 6 mercaptopurine	
#4 Melphalan	#4 3 years	#4 Local RT 4 years after diagnosis	#4 4 years	#4 White male Age 50	#4 Not stated	
#1 Melphalan	#1 1¼ year	#1 None	#1 1¼ years	#1 White male Age 57	#1 No response	—
#2 Melphalan Cyclophosphamide	#2 2¼ yrs 2 yrs	#2 Local to S I joint and R ilium	#2 4½ years	#2 White female Age 40	#2 No response	
#3 Cyclophosphamide Melphalan	#3 6 months 1¾ yrs	#3 To spine	#3 3 years	#3 White female Age 54	#3 Not stated but survived 1¾ yrs following diagnosis of leukemia	
#4 Melphalan	#4 3¼ years	#4 None	#4 3½ years	#4 White male Age 48	#4 Not stated but died 2 yrs after diagnosis of leukemia	
#5 Cyclophosphamide	#5 ¼ years	#5 None	#5 1¼ years	#5 Black female Age 78	#5 No response	
—	—	—	—	—	—	

Table 9-2 (continued)

Author; year pub; (yrs of study)	No. of cases	Leukemia type	No. of leukemia cases (% of cases observed)	Crude rate of ANLL per 1,000 person-years	Relative risk	Control population	Remission status at onset of leukemia
Gonzalez et al., 1977; (1965–73) [62]	476	ANLL	6 (1.3%) (5 developed sideroblastic anemia)	—	100 times expected for both ANLL and sidero-blastic anemia	Connecticut Tumor Registry	#1 — #2 — #3 — #4 — #5 — #6 —
Bergsagel et al., 1979 (1973–77) [63]	364	ANLL 10 AML 2 AMoL 2 Erythro-leukemia	14 (3.8%)	19	214	Canadian cancer incidence data	—
Kyle, 1982; (1965–75) [64]	908	16/17 AMMoL 1/17 Erythro-leukemia	17 (1.9%)	—	—	—	16/17 had improvement in myeloma at time of diagnosis of leukemia

[a]At the time of diagnosis of primary malignancy.
[b]First case.

Chemotherapy assoc. with leukemia	Duration of treatment	Role of radiation	Time between diagnosis of malignancy and diagnosis of leukemia	Age[a]/sex distribution of leukemia	Response of leukemia to treatment	Actuarial risk
#1 Melphalan, prednisone, procarbazine	—	#1 Given	#1 2½ years	#1 Female Age 58	#1 No response	—
#2 Melphalan, prednisone	—	#2 None	#2 6 years	#2 Male Age 65	#2 No response	
#3 Melphalan, prednisone	—	#3 None	#3 3 years	#3 Female Age 78	#3 Survived 8 months following diagnosis of anemia preceding leukemia	
#4 Melphalan, prednisone	—	#4 Given	#4 4½ years	#4 Male Age 59	#4 No response	
#5 Melphalan, prednisone, procarbazine	—	#5 None	#5 2 years	#5 Male Age 52	#5 No response	
#6 Melphalan, prednisone, procarbazine	—	#6 Given	#6 ½ year	#6 Male Age 67	#6 No response	
Melphalan, prednisone, cyclophosphamide, carmustine		Radiotherapy used for treatment of pain as needed		Age Group / No. 30–54: 1 55–64: 7 65–74: 5 ≥75: 1	—	17.4% at 50 months
8/17 had received daily oral melphalan 6/17 had intermittent oral melphalan 1/17 had intravenous melphalan 1/17 had BCNU 1/17 had cyclophosphamide	— — — — —	5/17 received radiotherapy	Median time was approximately 5 years	Age range was 40–79 Median was 56 years 12/17 men 5/17 women	Antileukemic treatment started for 4/17 cases but no remission. Median survival was 1 month for all cases.	For patients alive at 5 years the actuarial risk was 2.8%; for those alive at 10 years actuarial risk was 10.1%

with various modalities used in different schedules, making comparison difficult. Also, many of the investigators did not calculate either crude or age-adjusted rates of ANLL per person-years of observation following diagnoses of Hodgkin's disease nor are enough data presented to do this. Some of the studies did not employ a control population, making it impossible to compare observed rates of leukemia in the Hodgkin's cases with the expected rates in controls.

Relative risks (where measured) for leukemia as a secondary malignancy in Hodgkin's disease cases compared with controls range from 1 to 302. The crude incidence of leukemia per 1,000 person-years in Hodgkin's disease groups ranges from 0 to 25. Median interval from date of diagnosis of Hodgkin's disease to date of diagnosis of ANLL ranged from 2 years to 8 years. Of the 57 cases described in some detail in the tabular summary, 9 were treated with radiotherapy only, 8 with chemotherapy only, and 40 with both radiotherapy and chemotherapy.

Eleven analytic studies of groups of multiple myeloma cases are summarized in Table 9-2 [46, 55–64]. Similar problems to those described for the Hodgkin's disease studies characterize the follow-up studies of groups of multiple myeloma cases. Examination of the relative risks for development of ANLL following diagnosis (and, in most studies, treatment for multiple myeloma) shows that they range from 1 to 214. The only measure of ANLL incidence we could find was 19 per 1,000 person-years. Median time from diagnosis of multiple myeloma to diagnosis of ANLL ranged from 2½ to 6 years. Many of the cases (approximately 21–53) had been treated with melphalan only. Twelve of the 53 received melphalan plus radiotherapy; 1 received cyclophosphamide only, while 2 were treated with cyclophosphamide plus radiation; 1 received both melphalan and cyclophosphamide at different times in the course of treatment, while 2 received both of these agents and radiation; 14 received both melphalan and cyclophosphamide but it is unclear whether they had also been treated with radiotherapy. Most of the multiple myeloma cases were treated with radiation for relief of bony pain rather than for curative purposes in contrast to Hodgkin's disease.

While alkylating agents have clearly been shown to be carcinogenic in laboratory animals [8], the effects in humans, including possible leukemogenicity, are more difficult to interpret. Tables 9-1 and 9-2 appear to demonstrate an increased incidence of leukemia in Hodgkin's disease and multiple myeloma case groups compared with controls but the reason(s) for this increased risk have not been definitely established as a result of treatment with alkylating agents, radiotherapy, or a combination of these. It is difficult to examine the association between potentially carcinogenic agents and leukemia because of the improved survival brought about by these agents. ANLL may be part of the natural history of these hematopoietic disorders.

When left untreated, polycythemia vera, another hematopoietic disorder, is believed by many hematologists to spontaneously transform into ANLL in a low but significantly increased incidence of cases [65, 66]. This belief is based on a number of case reports but very few carefully conducted follow-up studies with appropriate comparison groups. An exception is a study carried out by Modan and Lilienfeld, who followed up 512 patients with polycythemia vera (of whom 79 were treated with x rays, 228 with P^{32}, 72 with x rays plus P^{32}, and 133 with no radiation). The polycythemia vera patients were compared with 476 patients with polycythemia of a questionable nature and 234 with polycythemia secondary to lung disease. Acute leukemia developed in 8.9% of the polycythemia vera patients treated with x rays, 11.0% of those treated with P^{32}, 16.7% of those treated with both x rays and P^{32}, and 0.8% of those treated with no radiation. In the secondary polycythemia group, 9.1% (one patient) of the 11 treated with x rays developed acute leukemia. There were no cases of acute leukemia in the 211 secondary polycythemia cases not treated with radiation [67].

A randomized controlled treatment trial of polycythemia vera patients compared treatment with phlebotomy alone to treatment with radioactive phosphorus and with chlorambucil (both agents supplemented by phlebotomy). The study showed that the risk of ANLL in patients given chlorambucil was 2.3 times that in patients given radioactive phosphorus and 13 times that in patients treated with phlebotomy alone [66]. No statistically significant differences in survival were found among the three treatment groups.

While these polycythemia vera data provide support for the theory that alkylating and/or radioactive agents increase the risk of ANLL in individuals with this disorder, another study of excess risk of ANLL (R.R. = 21) in patients with ovarian cancer (which rarely involves the bone marrow) treated with alkylating agents further supports the postulated leukemogenic character of these agents [18].

The increased risks of ANLL appear to be more than balanced by the improved survival due to use of alkylating agents and/or therapy in patients with Hodgkin's disease, ovarian cancer, and multiple myeloma. Most persons with these neoplasms, if untreated, would have a very short survival. The risks may not be justified in patients with polycythemia vera and those with non-neoplastic diseases, both of which may have a relatively long survival. This non-neoplastic group is of special concern because of the very short survival and poor prognosis associated with the ANLL when it does occur [68, 69].

While many case reports and case series [68–78] have described ANLL in persons with various non-neoplastic diseases (rheumatoid arthritis, psoriasis, multiple sclerosis) occurring after treatment with alkylating agents (including melphalan, chlorambucil, busulfan, cyclophosphamide), very few analytic studies have used case groups of adequate sample size with good follow-up for an appropriate length of time, comparison groups, and appropriate methods of analysis. One such follow-up study performed at two centers on 205 psoriasis patients treated with methotrexate found one case of AML developing when 0.13 were expected [79].

Hoover and Fraumeni examined the risk of malignancy in 6,297 kidney transplant recipients reported to the American College of Surgeons' Human Renal Transplant Registry during 1951–71 who had been followed for at least one month. No excess of leukemia was noted although the occurrence of reticulum cell sarcoma was 350 times expected [80].

REFERENCES

1. C.G. Moertel and A.B. Hagedorn, "Leukemia or lymphoma and coexistent primary malignant lesions: a review of the literature and a study of 120 cases," *Blood* **12:**788–802 (1957).
2. T. Whipham, "Splenic leukemia with carcinoma," *Tr. Path. Soc. London* **29:**313–321 (1878).
3. K. Borum, "Increasing frequency of acute myeloid leukemia complicating Hodgkin's disease: a review," *Cancer* **46:**1247–1252 (1980).
4. F. Rosner and H. Grünwald, "Hodgkin's discasc and acute leukemia," *Am. J. Med.* **58:**339–353 (1975).
5. R.S. Brody and D. Schottenfeld, "Multiple primary cancers in Hodgkin's disease," *Semin. Oncol.* **7:**187–201 (1980).
6. R.S. Brody, D. Schottenfeld, and A. Reid, "Multiple primary cancer risk after therapy for Hodgkin's disease," *Cancer* **40:**1917–1926 (1977).
7. S.M. Sieber, "The action of antitumor agents: a double-edged sword?" *Med. Pediatr. Oncology* **3:**123–131 (1977).
8. S.M. Sieber and R.H. Adamson, "The clastogenic, mutagenic, teratogenic and carcinogenic effects of various antineoplastic agents," in *Pharmacological Basis of Cancer Chemotherapy,* pp. 401–468, Twenty-Seventh Annual Symposium on Fundamental Cancer Research, 1974. Baltimore: Williams and Wilkins Co., 1975.
9. A.C. Griffin, E.L. Branat, and E.L. Tatum, "Nitrogen mustards as cancer inducing agents," *J. Am. Med. Assoc.* **144:**571 (1949).
10. J.H. Weisburger, D.P. Griswald, J.D. Prejean, A.E. Casey, H.B. Wood, and E.K. Weisburger, "The carcinogenic properties of some of

the principle drugs used in clinical cancer chemotherapy," *Recent Results Cancer Res.* **52**:1–17 (1975).

11. S.E. Walker and G.G. Bole, "Augmented incidence of neoplasia in female New Zealand black/New Zealand white (NZB/NZW) mice treated with long-term cyclophosphamide," *J. Lab. Clin. Med.* **78**:978–979 (1971).
12. D. Svoboda, J. Reddy, and C. Harris, "Invasive tumors induced in rats with actinomycin D," *Cancer Res.* **30**:2271–2279 (1970).
13. C. Bertazzoli, T. Chieli, and E. Solcia, "Different incidence of breast carcinoma or fibroadenomas in daunomycin or adriamycin treated rats," *Experientia* **27**:1209–1210 (1971).
14. T.P. Casey, "Azathioprine (Imuran) administration and the development of malignant lymphomas in NZB mice," *Clin. Exp. Immunol.* **3**:305–312 (1968).
15. F. Rosner and H. Grünwald, "Multiple myeloma terminality in acute leukemia—report of 12 cases and review of the literature," *Am. J. Med.* **57**:927–939 (1974).
16. J.F. Holland, "Epidemic acute leukemia," *N. Engl. J. Med.* **283**:1165–1166 (1970).
17. N. Einhorn, "Acute leukemia after chemotherapy (melphalan)," *Cancer* **41**:444–447 (1978).
18. R.R. Reimer, R. Hoover, J.F. Fraumeni, Jr. and R.C. Young, "Acute leukemia after alkylating-agent therapy of ovarian cancer," *N. Engl. J. Med.* **297**:177–180 (1977).
19. J.D. Boice, M.H. Greene, B.E. Greer, J.A. Blessing, A.J. Dembo, and R.A. Kleinerman, "Leukemia following alkylating agent therapy for ovarian cancer," *Am. J. Epidemiol.* (abs) **114**:440 (1981).
20. F. Rosner, R.W. Carey, and M.H. Zarrabi, "Breast cancer and acute leukemia: report of 24 cases and review of the literature," *Am. J. Hematol.* **4**:151–168 (1978).
21. D.H. Garfield, "Acute erythromegakaryocytic leukemia after treatment with cytostatic agents," *Lancet* (Letter to the Editor) **2**:1037 (1970).
22. E.C. Cadman, R.L. Capizzi, and J.R. Bertino, "Acute non-lymphocytic leukemia—a delayed complication of Hodgkin's disease therapy: analysis of 109 cases," *Cancer* **40**:1280–1296 (1977).
23. F. Rosner and H.W. Grünwald, "Multiple myeloma and Waldenström's macroglobulinemia terminating in acute leukemia," *N.Y. State Med. J.* **80**:558–570 (1980).
24. M.H. Zarrabi, F. Rosner, and J.M. Bennett, "Non-Hodgkin's lymphoma and acute myeloblastic leukemia: report of 12 cases and review of the literature," *Cancer* **44**:1070–1080 (1979).
25. M.H. Zarrabi, H.W. Grünwald, and F. Rosner, "Chronic lympho-

cytic leukemia terminating in acute leukemia," *Arch. Intern. Med.* **137**:1059–1064 (1977).

26. J.H. Hornbaker, "Giant follicular lymphoblastoma terminating in acute myelogenous leukemia," *Ann. Intern. Med.* **53**:221–227 (1960).
27. J.L. Poth, R.P. George, W.P. Creger, and S.L. Schrier, "Acute myelogenous leukemia following localized radiotherapy," *Arch. Intern. Med.* **128**:802–805 (1971).
28. R.B. Weiss, R.D. Brunning, and B.J. Kennedy, "Lymphosarcoma terminating in acute myelogenous leukemia," *Cancer* **30**:1275–1278 (1972).
29. F.W. Gunz, J.A. Levi, D.E. Lind, and P.C. Vincent, "Development of acute leukemia in a patient with lymphosarcoma," *N.Z. Med. J.* **78**:71–75 (1973).
30. W.S.A. Allan, "Acute myeloid leukemia after treatment with cytostatic agents," *Lancet* (Letter to the Editor) **2**:775 (1970).
31. C.G.S. Smit and L. Meyler, "Acute myeloid leukemia after treatment with cytostatic agents," *Lancet* **2**:671–672 (1970).
32. R.A. Kaslow, N. Wisch, and J.L. Glass, "Acute leukemia following cytostatic chemotherapy," *J. Am. Med. Assoc.* **219**:75–76 (1972).
33. E.M. Greenspan and B.G. Tung, "Acute myeloblastic leukemia after cure of ovarian cancer," *J. Am. Med. Assoc.* **230**:418–420 (1974).
34. G. Sotrel, K. Jafari, A.F. Lash, and R.C. Stepto, "Acute leukemia in advanced ovarian carcinoma after treatment with alkylating agents," *Obstet. Gynecol.* **47** *Suppl.:* 67S–71S (1976).
35. R.B. Solomon and D. Firat, "Acute leukemia following treatment with irradiation and alkylating agent," *N.Y. State J. Med.* **71**:2422–2425 (1971).
36. R.J. Cohen, P.H. Wiernik, and M.D. Walker, "Acute non-lymphocytic leukemia associated with nitrosourea chemotherapy: report of two cases," *Cancer Treat. Rep.* **60**:1257–1261 (1976).
37. D.A. Casciato and J.L. Scott, "Acute leukemia following prolonged cytotoxic agent therapy," *Medicine* **58**:32–47 (1979).
38. M.V. Peters and K.C.H. Middlemiss, "A study of Hodgkin's disease treated by irradiation," *Am. J. Roentgenol. Radium Ther. Nucl. Med.* **79**:114–121 (1958).
39. J.W. Berg, "The incidence of multiple primary cancers. I. The development of further cancers in patients with lymphomas, leukemias and myeloma," *J. Natl. Cancer Inst.* **38**:741–752 (1967).
40. E.Z. Ezdinli, J.E. Sokal, C.W. Aungst, U. Kim, and A.A. Sandberg, "Myeloid leukemia in Hodgkin's disease: chromosomal abnormalities," *Ann. Intern. Med.* **71**:1097–1104 (1969).
41. W.H. Crosby, "Acute granulocytic leukemia, a complication of therapy in Hodgkin's disease?" *Clin. Res.* **17**:463 (1969).

42. D.R. Newman, J.E. Maldonado, E.G. Hamson, J.M. Kiely, and J.W. Linman, "Myelomonocytic leukemia in Hodgkin's disease," *Cancer* **25:**128–134 (1970).
43. P.L. Weiden, K.G. Lerner, A. Gerdes, J.D. Heywood, A. Fefer, and E.D. Thomas, "Pancytopenia and leukemia in Hodgkin's disease: report of three cases," *Blood* **42:**571–577 (1973).
44. G. Bonadonna, M. De Lena, A. Banfi, and A. Lattuada, "Secondary neoplasms in malignant lymphomas after intensive therapy," *N. Engl. J. Med.* (Letter to the Editor) **288:**1242–1243 (1973).
45. G.J. Sahakian, H. Al-Mondhiry, M.J. Lacher, and C.E. Connolly, "Acute leukemia in Hodgkin's disease," *Cancer* **33:**1369–1375 (1974).
46. G.R. Newell, E.T. Krementz, J.D. Roberts, and B.K. Kinnear, "Multiple primary neoplasms in blacks compared to whites. I. Further cancers in patients with Hodgkin's disease, leukemia and myeloma," *J. Natl. Cancer Inst.* **52:**635–638 (1974).
47. G.P. Canellos, V.T. DeVita, J.C. Arseneau, J. Whang-Peng, and R.E. Johnson, "Second malignancies complicating Hodgkin's disease in remission," *Lancet* **1:**947–949 (1975).
48. C. Jacquillat, G. Auclerc, M. Weil, M. Boiron, and J. Bernard, "Acute leukemia, Kaposi's sarcoma, epitheliomas, complicating 30 observations of Hodgkin's disease," *Proc. Am. Assoc. Cancer Res.* **17:**247 (1976).
49. J. Larsen and H. Brincker, "The incidence and characteristics of acute myeloid leukaemia arising in Hodgkin's disease," *Scand. J. Haematol.* **18:**197–206 (1977).
50. G.R. Newell and E.T. Krementz, "Multiple malignant neoplasms in the Charity Hospital of Louisana Tumor Registry," *Cancer* **40:**1812–1820 (1977).
51. C.N. Coleman, C.J. Williams, A. Flint, E.J. Glatstein, S.A. Rosenberg, and H.S. Kaplan, "Hematologic neoplasia in patients treated for Hodgkin's disease," *N. Engl. J. Med.* **297:**1249–1252 (1977).
52. D.M. Toland, C.A. Coltman, and T.E. Moon, "Second malignancies complicating Hodgkin's disease: The Southwest Oncology Group experience," *Cancer Clin. Trials* **1:**27–33 (1978).
53. P. Valagussa, A. Santoro, R. Kenda, F. Fossati Bellani, F. Franchi, A. Banfi, F. Rilke, and G. Bonadonna, "Second malignancies in Hodgkin's disease: a complication of certain forms of treatment," *Br. Med. J.* **280:**216–219 (1980).
54. M. Baccarani, A. Bosi, and G. Papa, "Second malignancy in patients treated for Hodgkin's disease," *Cancer* **46:**1735–1740 (1980).
55. R.A. Weitzel, "Carcinoma co-existent with malignant disorders of plasma cells," *Cancer* **11:**546–549 (1958).

56. N.G. Nordenson, "Myelomatosis. A clinical review of 310 cases," *Acta Med. Scand. Suppl.* **445**:178–186 (1966).
57. G.A. Edwards and Z.A. Zawadzki, "Extraosseous lesions in plasma cell myeloma," *Am. J. Med.* **43**:194–205 (1967).
58. R.A. Kyle, R.V. Pierre, and E.D. Bayrd, "Multiple myeloma and acute myelomonocytic leukemia," *N. Engl. J. Med.* **283**:1121–1125 (1970).
59. E. Anderson and A. Videback, "Stem cell leukaemia in myelomatosis," *Scand. J. Haematol.* **7**:201–207 (1970).
60. M. Khaleeli, W.M. Keane, and G.R. Lee, "Sideroblastic anemia in multiple myeloma: a preleukemic change," *Blood* **41**:17–25 (1973).
61. R.K. Karchmer, M. Amare, W.E. Larsen, A.G. Mallouk, and G.G. Caldwell, "Alkylating agents as leukemogens in multiple myeloma," *Cancer* **33**:1103–1107 (1974).
62. F. Gonzalez, J.M. Trujillo, and R. Alexanian, "Acute leukemia in multiple myeloma," *Ann. Intern. Med.* **86**:440–443 (1977).
63. D.E. Bergsagel, A.J. Bailey, G.R. Langley, R.N. MacDonald, D.F. White, and A.B. Miller, "The chemotherapy of plasma-cell myeloma and the incidence of acute leukemia," *N. Engl. J. Med.* **301**:743–748 (1979).
64. R.A. Kyle, "Second malignancies associated with chemotherapeutic agents," *Semin. Oncol.* **9**:131–142 (1982).
65. S.A. Landow, "Acute leukemia in polycythemia vera," *Semin. Hematol.* **13**:33–48 (1976).
66. P.D. Berk, J.D. Goldberg, M.N. Silverstein, A. Weinfeld, P.B. Donovan, J.T. Ellis, S.A. Landaw, J. Laszlo, Y. Najean, A.V. Pisciotta, and L.R. Wasserman, "Increased incidence of acute leukemia in polycythemia vera associated with chlorambucil therapy," *N. Engl. J. Med.* **304**:441–447 (1981).
67. B. Modan and A.M. Lilienfeld, "Polycythemia vera and leukemia—the role of radiation treatment: a study of 1,222 patients," *Medicine* **44**:305–344 (1965).
68. H.W. Grünwald and F. Rosner, "Acute leukemia and immunosuppressive drug use," *Arch. Intern. Med.* **139**:461–466 (1979).
69. J.P. Aymard, J. Frustin, F. Witz, J.N. Colomb, P. Lederlin, and R. Herbeuval, "Acute leukemia after prolonged chlorambucil treatment for non-malignant disease. Report of a case and literature survey," *Acta Haematol.* **63**:283–285 (1980).
70. R.B. Rees, J.H. Bennett, E.M. Hamlin, and H.I. Maibach, "Aminopterin for psoriasis: a decade's observation," *Arch. Dermatol.* **90**:544–552 (1964).
71. P. Stavem and M. Harboe, "Acute erythroleukemia in a patient

treated with melphalan for the cold agglutinin syndrome," *Scand. J. Haematol.* **8:**375–379 (1971).

72. B.H. Pollack, J.H. Barr, Jr., B.L. Stolzer, C.H. Eisenbeis, A. Agarwal, and H.M. Margolis, "Neoplasia and cyclophosphamide," *Arthritis Rheum.* (Letter to the Editor) **16:**524–526 (1973).
73. C.D. Coban, R.P. Sheon, and A.B. Kirsner, "Immunosuppressive drugs and acute leukemia," *Ann. Intern. Med.* **79:**131–132 (1973).
74. A.J. Silvergleid and S.L. Shrier, "Acute myelogenous leukemia in two patients treated with azothioprine for non-malignant diseases," *Am. J. Med.* **57:**885–888 (1974).
75. R.R. Love and J.M. Sowa, "Myelomonocytic leukaemia following cyclophosphamide therapy of rheumatoid disease," *Ann. Rheum. Dis.* **34:**534–535 (1975).
76. E. Alexson and K.D. Brandt, "Leukaemia after azathioprine treatment of connective tissue disease," *Am. J. Med. Sci.* **273:**335–340 (1977).
77. R.M. Bukowski, J.K. Weick, R.R. Reimer, G.W. Groppe, and J.S. Hewlett, "Characteristics of acute leukemia in patients with non-malignant diseases receiving alkylating agent therapy," *Blood* **50** *Suppl.:*185 (1977).
78. C. Kleinknecht, P. Guesry, G. Lenoir, and M. Broyer, "High cost benefit of chlorambucil in frequently relapsing nephrosis," *N. Engl. J. Med.* (Letter to the Editor) **296:**48 (1977).
79. P.L. Bailin, J.P. Tindall, H.H. Roenigk, and M.D. Hogan, "Is methotrexate therapy for psoriasis carcinogenic?" *J. Am. Med. Assoc.* **232:**359–362 (1975).
80. R. Hoover and J.F. Fraumeni, Jr., "Risk of cancer in renal transplant recipients," *Lancet* **2:**55–57 (1973).

10 An Overview

It is clear that leukemias consist of several diverse, etiologically distinct diseases. Recent advances in the understanding of cytochemical, cytogenetic, and immunologic marker differences between the leukemias as well as increasingly successful treatment approaches have made it difficult over the past few years to develop a single classification scheme with long-term usefulness. Information that has been obtained from epidemiologic studies that considered all leukemias a single disease is also less helpful in understanding etiology. Although a number of studies in the past 10–20 years have used the histopathologic classification proposed in this review as a compromise between older and newer schemes, a number of investigators still continue to consider all of the leukemias as a single entity.

In spite of a large body of research investigating pathogenesis and etiology, risk factors that have been established for each type of leukemia account for only a very small proportion of cases. Nevertheless, it is probably useful to summarize known and suspected risk factors for each and to identify promising areas for further research.

ALL

The bimodal age distribution for this leukemia, with an early childhood peak, then an antimode in young and middle adult years followed by a rising incidence in the elderly, suggests that different risk factors may be operative at the different ages. It is not clear from studies examining secular trends whether ALL has actually increased in incidence during the twentieth century or whether the apparent increase is entirely due to changes in classification. A large proportion of cases, particularly children, formerly classified as unspecified or undifferentiated in more recent years are now being classified as having ALL.

The significantly lower incidence and ratio of childhood ALL to ANLL in Asians and Africans compared with that for western nations suggests that potential differences in genetic characteristics, lifestyle factors, and exposure to infectious agents and to environmental factors deserve additional study. HLA, chromosome, immunologic, and other appropriate studies comparing Asians and Africans with westerners would be worthwhile in the context of population-based epidemiologic studies comparing infectious exposures, lifestyle, and environmental factors between these groups. Additional support for the role of genetic factors in childhood ALL comes from twin studies, most of which were carried out 10–15 years ago and did not specify cell type. Nevertheless, the 3:1 ratio for ALL:ANLL in the U.S. and Great Britain, where most of the twin studies have been carried out, makes it likely that ALL accounted for 25% concordance in monozygotic twins. Increased incidence of ALL in persons with congenital and hereditary disorders characterized by inherited or early-in-life induced chromosome abnormalities (such as Down's, Klinefelter's, Bloom's syndrome, or Fanconi's anemia) is additional evidence for the role of genetic factors. Since a large proportion of such persons do not develop ALL, it is likely that leukemia results from the interaction of chromosomal abnormalities and environmental factors. Although the numbers of such persons are very small, genetic–epidemiologic studies of these groups and their families may be very fruitful.

The immunologic system also seems to be involved in the development of ALL, which appears to occur in excess in some immunodeficiency states. The levels of risk, however, have not been quantified for groups with immunodeficiency diseases. There are several studies providing weak and somewhat equivocal evidence that ALL may also be decreased in incidence in children who received BCG immunization neonatally.

Both high-dose and low-dose radiation appear to increase the risk for ALL in childhood. The latent period seems to vary with age at exposure. Radiation is one type of exposure for which current knowledge is sufficient to suggest that any exposure level ought to be minimized. In contrast, the significance of data from occupational studies showing associations between manufacturing and sales with ALL in males is not clear.

Evidence is suggestive that pregnant women who develop varicella and influenza in pregnancy are at increased risk for having offspring who develop ALL. This again suggests that the immune system (of the particularly susceptible fetus) may play a role in the development of ALL. Although a viral etiology has not yet been proven to cause childhood ALL, the recent discovery of a new rotavirus causing some cases of human T-cell lymphosarcoma cell leukemia (the human T-cell leukemia/lymphoma virus or HTLV) suggests that is is possible that a previously unknown leukemogenic virus might also be found for ALL.

Most of the epidemiologic data for ALL pertains to childhood ALL. Very little is known about ALL occurring at older ages or why young and middle-aged adults have a relatively low frequency. It is possible that immunologic characteristics of the young and the elderly may be very important in the etiology of ALL. Additional study of immunologic factors within the context of case-control studies for both the young and the elderly with ALL seems warranted.

CLL

Like ALL, CLL also appears to have a strong immunogenetic component. CLL is primarily confined to the late middle-aged and the elderly. It is very rare in Asians and remains rare in Asian migrants to western nations.

An indication that genetic factors may be important is the strong familial component observed in CLL. CLL or other lymphoproliferative malignancy has been found, in a number of studies, in 8%–10% of first-degree family members of cases. Of course, interactions with environmental factors must be considered in interpreting such findings. But the age distribution of persons with CLL is such that affected individuals are likely to have been separated from first-degree relatives for 30–40 years prior to onset of CLL. Hence any shared environmental factors must either continue to be experienced by affected relatives after they move away from the family home or there is a long latent period between exposure(s) and the development of CLL or other lymphoproliferative malignancy.

The role of immunologic factors also appears to be important. A number of case series and case-control studies suggest that CLL may be associated with chronic antigenic stimulation from a variety of chronic inflammatory and/or autoimmune diseases. Epidemiologic population-based case-control and family studies exploring interactions between genetic, immunologic, and environmental factors could be instructive in elucidating etiology.

CLL appears to be increased in incidence in rural areas and in farmers. The exact exposure(s) responsible have not been identified but there is some evidence that certain types of farming as well as proximity to cows, poultry, and other animals susceptible to virally caused leukemia are particularly associated with human cases. Additional epidemiologic studies of high risk groups of farmers could be helpful in identifying specific leukemogenic exposures.

Exposure to certain solvents also appears to increase the risk for CLL. The precise types of solvents have not been completely identified but benzene appears to be one. CLL has also been found to have a higher incidence in rubber workers. Other occupational associations include construction, manufacturing, utilities, and communication industries for males and other personal services, professional and related services, and medical services for females only. The specific exposures that increase the risk in each of these industries have not been identified.

Finally, the interesting trend of apparent increasing incidence of CLL in the elderly as well as the increasing incidence in males ought to continue to be studied.

ANLL

This leukemia type appears to have an increase in incidence in many countries over the last few decades for both adults and children, particularly in urban, industrial areas. Most childhood leukemias in Asians are due to ANLL. And yet most other evidence suggests that genetic and immunologic factors play a less important role in this type of leukemia than in ALL or CLL.

Exposure to radiation appears to cause increased incidence of ANLL. Of interest is the deficit of ANLL and excess of CML in Hiroshima compared with Nagasaki, believed to be due to differences between the types of radiation released by the two bombs. This is currently being studied.

ANLL is increased in a number of industries in which there is substantial benzene exposure, including the shoe leather, pliofilm, and chemical industries. The increasing incidence of this leukemia in industrial areas and the known association with benzene and possibly other benzene-containing chemicals or derivatives suggests that studies of ANLL should focus more sharply on chemical exposures.

ANLL is also increased in incidence in patient populations treated with alkylating and other chemotherapy drugs. Persons with lymphoproliferative malignancies (Hodgkin's disease and multiple myeloma) treated with such therapies have been repeatedly confirmed as having excess ANLL as have populations with solid tumors (ovarian) treated with alkylating agents. Although this accounts for a small proportion of ANLL, it must be kept in mind when determining overall survival and cure rates for these neoplasms. Of greater concern ethically are groups of patients with nonmalignant, mostly autoimmune disorders (such as rheumatoid arthritis, psoriasis) who are treated with alkylating and other leukemogenic chemotherapy agents. This appears to be a subgroup of ANLL, albeit small, which is preventable. Additional epidemiologic studies of drugs, purported to be leukemogenic, including phenylbutazone and chloramphenicol, are needed before the nature of any postulated association can be assessed.

CML

This is the least understood of the leukemias in terms of etiology. Genetic and immunologic factors do not appear to be important. There is very little variation in incidence worldwide or regionally. The intriguing occurrence of the acquired Ph^1 chromosome in 90% of cases suggests that agents that cause chromosomal breaks deserve additional attention. These include ionizing radiation, known to be associated with CML, and chemicals.

Despite great advances in many scientific disciplines in which in-depth study has focused upon leukemia pathogenesis and etiology, the risk factors identified account for only a small proportion of each of the leukemias. There are a number of promising areas for additional research. Yet, additional clues leading to important breakthroughs in identification of etiologic factors are likely to come from a combination of laboratory studies (tumor virology, chemical carcinogenesis study in animals) and multidisciplinary epidemiologic studies in which epidemiologists, geneticists, immunologists, virologists, toxicologists, etc., examine interactions between environmental, genetic, and immunologic factors.

Author Index

Note: Roman page numbers indicate text citations; page numbers in italics indicate reference list pages.

Subject Index